Praise for
The Art of Asking

"Terry's insights on this essential subject are brilliant and practical at the same time. His stories are wonderful and drive his points home with both humor and stark clarity. He provides questions and approaches that one can implement immediately. Required reading for every leader who wishes to see his or her organization flourish and career progress."

—Garry A. Neil, MD, Corporate Vice President, Johnson & Johnson

"Terry has done it. Asking, listening, understanding the real meaning of the answers, and taking actions based on facts are really the essence of managing a process, an organization, or a corporation. This book has helped me in connecting the dots in my understanding (and lack thereof) of why things really did not work the way I expected them to work. *The Art of Asking* is really a practical guide for everyday management. This book should be a part of any core curriculum for management training, irrespective of the managers' areas of focus."

—Pradip Banerjee, PhD, Chairman and Chief Executive Officer, Xybion; retired partner, Accenture

"This is a unique and valuable guide to asking the right question at the right time. It is also about using insightful questions to exercise leadership. As Peter Drucker once observed, 'The leader of the past knew how to tell. The leader of the future will know how to ask.'"

—George Day, Geoffrey T. Bosi Professor of Marketing, Wharton Business School

"The framework and techniques provide outstanding ideas for executives to both gain better information and develop the analytical skills of their teams through Socratic learning."

—Terry Hisey, Vice Chairman and U.S. Life Sciences Leader, Deloitte

"An easy read! Terry shares his past corporate experiences and explains in simple terms how to improve your chances of success through better questioning. Right on the money!"

—Mark Hopkins, entrepreneur and small business owner

THE ART OF ASKING

ASKING

ASK BETTER QUESTIONS, GET BETTER ANSWERS

TERRY J. FADEM

Vice President, Publisher: Tim Moore
Associate Publisher and Director of Marketing: Amy Neidlinger
Acquisitions Editor: Jennifer Simon
Editorial Assistants: Myesha Graham and Heather Luciano
Development Editor: Russ Hall
Operations Manager: Gina Kanouse
Digital Marketing Manager: Julie Phifer
Publicity Manager: Laura Czaja
Assistant Marketing Manager: Megan Colvin
Managing Editor: Kristy Hart
Project Editor: Betsy Harris
Copy Editor: Keith Cline
Proofreader: San Dee Phillips
Interior Designer: Gloria Schurick
Cover Designer: Tobias Design
Compositor: Nonie Ratcliff
Manufacturing Buyer: Dan Uhrig

Publishing as FT Press
Upper Saddle River, New Jersey 07458

FT Press offers excellent discounts on this book when ordered in quantity for bulk purchases or special sales. For more information, please contact U.S. Corporate and Government Sales, 1-800-382-3419, corpsales@pearsontechgroup.com. For sales outside the U.S., please contact International Sales at international@pearsoned.com.

Company and product names mentioned herein are the trademarks or registered trademarks of their respective owners.

Printed in the United States of America

First Printing December 2008

ISBN-10: 0-13-714424-5
ISBN-13: 978-0-13-714424-2

Pearson Education LTD.
Pearson Education Australia PTY, Limited.
Pearson Education Singapore, Pte. Ltd.
Pearson Education North Asia, Ltd.
Pearson Education Canada, Ltd.
Pearson Educatión de Mexico, S.A. de C.V.
Pearson Education—Japan
Pearson Education Malaysia, Pte. Ltd.

Library of Congress Cataloging-in-Publication Data

Fadem, T. J. (Terry Jay), 1948-
   The art of asking : ask better questions, get better answers / T.J Fadem.
      p. cm.
   Includes bibliographical references.
   ISBN 0-13-714424-5 (pbk. : alk. paper) 1. Communication in management. 2. Management.
3. Executive ability. 4. Organizational behavior. I. Title.
   HD30.3.F33 2008
   658.4'5—dc22

2008007868

Contents

Contents

Contents

But dost thou know what will be tomorrow?

—*The Grand Inquisitor, in* The Grand Inquisitor,
by Fyodor Dostoevsky

Disclaimer

The events described in this book are true to the best of the author's ability to recall them. Names and details that might identify people or companies have been changed, wherever possible, to protect the identity of the innocent as well as the guilty.

Acknowledgments

I want to acknowledge the contributions of my family—Susan, Lynne, and Charles Fadem—for their patience in listening to my rendition of the stories that appear on the following pages, over and over again. And to Q and Ginger, who both barked furiously at me to walk them whenever I was deep in thought about some important management questions, as if to remind me that good management is often something that just happens.

About the Author

T.J. (Terry) Fadem is a veteran manager with 25 years of experience ranging from supervising steel workers (J&L Steel) to managing in a major corporation (DuPont) to working with start-up companies. His business venture teams have been profiled in books and periodicals, and he has also been a frequent speaker and consultant on strategic management issues. Fadem is currently the managing director, Corporate Alliances at the School of Medicine at the University of Pennsylvania where he is also a member of the Core Team of the Mack Center for Technological Innovation at the Wharton School. In addition, Fadem is president of the Biomedical Research and Education Foundation.

Preface:
Corporate Inquisitions

The Not-So-Grand Inquisitor

He sat behind a desk holding a pitchfork, as all upper-level managers do when they are bedeviling their employees. Well, at least that's how many people picture the boss. This guy was actually holding one that resembled an implement of the devil—a long black handle with a red trident at the end. An appropriate accompaniment for a kid in a Halloween costume, it was out of place with the corporate blue suit of the middle-aged business director. But, as you will see in a moment, this was a manager who was out of step with his business.

Managing business development efforts for a major electronics company with worldwide supply chains is a daunting job for anyone. He had responsibility for overseeing a major growth initiative for the company—one that would likely determine the future of the division. His infrequent visits to the facility that housed the main business unit were as welcome as the arrival of bird flu.

The project he had come to review was beset with problems. The marketing organization criticized research, believing that the product design would not meet customer expectations. Every redesign that satisfied the demands of marketing added costs to the product that threatened to price it out of the market it was targeted to reach, thus making the salespeople very unhappy. And they all argued with manufacturing because no matter what design was settled on, no one in the plant had any confidence the product could be manufactured reliably. The project was woefully behind schedule and so far over budget that the likelihood of recovering development costs had become a major concern for management.

The desk that he was using, as a physical barrier between himself and the team as much as anything else, rested on a concrete platform, about a foot high, in an old factory warehouse. He actually needed the desk because this particular group of employees was known to hurl chairs when they disagreed with each other. Who knows what they might heave at this guy? Although the original purpose of the elevated floor was to keep gunpowder dry, it now functioned as a stage on which the manager attempted to transform himself into an inquisitor. He had come to visit a business team that was producing a seemingly unending stream of problems rather than products.

This meeting was convened to find solutions to the problems the development team had been having so that production could be scheduled and the sales force could start to take orders. In reality, the director was holding an inquisition. He believed he knew the answers—he just wanted to ask the questions. So, with evil scepter in hand, he conducted an investigation, calling on his victims by pointing his pitchfork at them as if to skewer each respondent on one of the barbs.

He pointed to the engineering supervisor.

> **Inquisitor:** What do you mean you can't get the boards to work? Who designed them? Who built the prototypes?

He paused here to catch his breath. No one was going to speak.

> **Inquisitor:** What's wrong with you people? Can you explain this?

The silence continued. There was no answer to his bullying, except for one of the engineers who entered the room late. A particularly brilliant designer, "Doctor Doom" accepted the verbal challenge.

> **Inquisitor:** The project is now overdue by six months. Not one part can be produced for the original forecasted cost. What is the final projected cost of the production model now?
>
> **Dr. Doom:** About four times what we planned at the start!

Delivering bad new was Dr. Doom's specialty, hence the nickname. He appeared to enjoy telling managers the truth, as he saw it, and seemed especially pleased if it was very bad news that was not expected by management—and this was indeed, very bad news.

Inquisitor: We're losing time, and now you tell me that we have lost any hope of having a price advantage? Do any of you think this makes sense? Can you explain this?

He pointed to one of the marketing people in the room.

Inquisitor: What's wrong with you people? How could you let this happen?

Although offensive by most standards, this particular inquisitor was a small *i* inquisitor in this company. He was threatening to lower-level people, but in reality, he had only a limited ability to dismiss staff or end careers. This company had an abundance of inquisitors in training.

The business director worked for a general manager who was the real "Grand Inquisitor" of the company. The GM was so good at inquisitions that careers spontaneously combusted under the intense pressure of his examinations during business review meetings. There was no need for any burning at the stake. He was known to extract resignations on the spot.

· ✍ ·

If you work in a company or any organization long enough, you might eventually attend or participate in an inquisition or two. I have seen a number of them, and I believe most if not all inquisitions are unnecessary. Although hopefully not commonplace, they do happen, and they represent many of the worst characteristics of inappropriate questioning conducted by managers in their daily work. One of the reasons that unsuitable questioning occurs is that the skills employed when conducting inquiries tend to be those that are passed on by example.

If mentors or senior managers are particularly good at asking questions (and if they are also personally successful), their skills are passed on to those who want to emulate them. As in the case just discussed, however, if managers' skills are tactless and the company is still successful under this kind of leadership, the reverse happens. Poor habits are perpetuated. People are fooled into thinking that bullying, intimidating, or torturing by "elocutioning" employees can bring success just because they see these traits in managers of successful enterprises. Even when a business fails, if people had no other mentors to learn from, they have few positive skills to take forward in their career.

Unfortunately, as discussed in the next few chapters, not all successful managers, including those who possess excellent questioning skills, excel at asking questions all the time.

Managers ask questions for a wide variety of specific reasons. For purposes of our discussion, I have boiled the reasons down to three general categories of inquiry:

1. Questions asked because the answer is important
2. Questions asked because the question is important
3. Questions asked because the process of asking is important

In the first category, the answer is more important than the question, so all questions need to be asked with that fact in mind. A manager might want to learn about an idea, or, as in the previously discussed case, the issue under investigation may be "what went wrong" (even though the management devil in this example had no real interest in the answer). In addition to asking questions effectively, managers need simultaneously to employ listening skills.

In the second category, the question is more important than the answer. A manager might want a particular line of reasoning to be used to evaluate projects, or perhaps other considerations should be addressed and the question is a tool to be applied to the situation. There might be no answer to some questions because they are designed to generate discussion rather than answers. This practice is common in many classrooms where questions are designed to get students to think about the question or sharpen their analytical skills rather than supply a correct answer. The business director in the short example mentioned previously actually didn't care about the questions either.

For him, the process of asking questions—the inquisition—was what mattered most. So, the manner of asking seemed to be his overriding concern. He was intentionally making people feel uncomfortable by grilling them with questions and letting them know that he didn't care about the answers. But, there are less-threatening circumstances where the process of asking is designed by the manager to instruct or to get a group or an individual to approach a problem differently. Mentors, professors, consultants, and advisors often play this role with their questions.

In other cases, the process of asking can be used to allow the group to develop new ideas. So, not all questioning that focuses on process is an inquisition.

For most managers, interest lies in the question as well as the answer. And, it's the process of asking that establishes the importance of each question and answer. This book was designed to address the need for improving questions as well as the manner in which managers ask them—to thus achieve better answers, which are ultimately what is needed for a business to succeed in the long run.

· ✿ ·

The business management "devil" in the opening story eventually met his due. After months of delays, the development team finally produced a viable product, over a year behind schedule and well over budget—but it was completed. Fighting stopped long enough to get the job done, and everyone on the project team moved on to other assignments or left for jobs elsewhere.

By sheer coincidence, I happened to be present when this particular business director was making a presentation to the Grand Inquisitor—the general manager of the division.

The director's business unit was not doing well. Errors in judgment coupled with poor forecasting had led to two straight years of underperformance. New products were delayed, morale was poor, and there was no end in sight. Under intense questioning from the general manager, he self-destructed.

> **Grand Inquisitor:** Forget the numbers, what's your analysis of the situation? Why do you have such bad news about something you should've fixed long before now?
>
> **Director:** Exchange rates hurt our European margins, costs are up in our plants in Asia due to environmental concerns, and marketing forecasted a more aggressive return than the sales force was able to deliver.
>
> **Grand Inquisitor:** That's not good enough. And what else?
>
> **Director:** Well, if my analysis isn't good enough for you, you can find someone else to run this business.
>
> **Grand Inquisitor** (rather triumphantly): I will.

And then, in a perverse twist of the pitchfork, which was now buried deep in the manager's ego, he gleefully moved on to attack any self-esteem that remained.

> **Grand Inquisitor:** But I'm still waiting for a good explanation from you for such a bad performance.

Turning red and gasping for air, the exasperated manager stalked out of the meeting and immediately resigned.

The business eventually failed and was sold off by the parent company. It's now doing better under different management. I lost track of the not-so-grand inquisitor, but the Grand Inquisitor general manager went on to be an even more overbearing boor of a manager as the CEO of another company, where he was eventually awarded a golden parachute for leading yet another crop of business managers to extinction. There was little in his record of leading businesses that demonstrated the kind of good performance expected from a true business leader. A mythology followed him around with vague references to how he saved a business with his great insights fresh out of school with his MBA. I never met a witness to this history. He just always seemed to appear on the scene when the business conditions were failing just enough to blame previous management, impress people with excruciatingly tortuous questioning sessions, and then leave the business right before it crashed. To this day, I cannot understand why he was never held to account for years of poor performance, not to mention ruined careers. This lack of accountability is a troubling aspect of business management visible in a number of places in the market.

·◈·

Improving a manager's ability to ask a question is no guarantee that business performance will be improved, but it should help. It is possible that both the project and the business of the manager cited in the preceding example might have performed better if the manager had been better prepared to ask questions of the people who worked for him. This book offers an opportunity for managers at all levels to improve one of the most basic aspects of their jobs: asking questions!

Introduction: Questioning Is the Skill of Management

1. Is There a Basic Set of Management Questions?

Yes.

All managers can use a basic set of questions, at any level in any organization, in any situation, anywhere in the world, and in any language. These questions are tools that should be issued to each manager when he or she joins the profession. Most professionals have a basic set of implements to use in their craft. Carpenters have hammers, dentists have picks, and physicians have stethoscopes. It is hard to envision any of these people working in their chosen fields without their basic set of tools. Managers, too, have a basic set of tools: *questions*. And nothing is as simple, or as complex, for a manager, or for any person in any position of authority and responsibility, than asking questions.

Some of us are very good at it. We always seem to ask the right question at the right time. Others of us are less well prepared, and our questions often do not yield the kinds of results we want or that the business needs. Even the best among us are subject to a number of common errors. So, all of us managers share the need to improve our skills. Before we start discussing the details of common questioning errors, a quick review of the basic tools of management is provided.

If you want the basics, the following list should suffice. There is a lot more to asking a question than merely using an interrogative, but these words do cover the full managerial spectrum of interrogation.

Basic Questions

For all managers in any situation at any time

What?

Where?

When?

Why?

Who?

How?

How much?

What if?[1]

These questions are universally applicable. If you are ever in a situation where you need a question, or if you want to make certain all questions have been asked, just run down the list. This list also serves as a handy checklist when you need to make a quick decision. Consider which of the issues implied by these interrogative words have not been addressed in the situation you are facing, and then raise those issues.

This list happens to be my personal shorthand way of making certain I have covered all perspectives in a discussion. You can add a number of other questioning words and phrases. Words such as *which, is, could, would, should, do, can, will,* and so on are used every day and could be the basis for another list. It all depends on what you expect to accomplish.

Organizations need all their managers to be successful, not just the ones who ultimately end up in the executive suite. The purpose of focusing on improving the quality of questions is to improve the quality of management, all management.

Success does not necessarily follow the ability to ask questions. It rests on the confluence of a lot of variables. However, by spending time considering how to improve a basic management skill—questioning—the outcome should certainly be better than what it would have been otherwise.

Could any of the well-known corporate disasters of our day have been avoided with better questions, asked more often by more people, such as the boards of these companies? We will never know. However, by improving questioning skills among more managers in a business, chances are good that other disasters can be averted in the future.

You can use the monosyllabic queries previously listed, if you choose, or you can work on developing questioning as a skill. Either way, the purpose of this book is to influence managers to think about questioning differently. In addition to the basic list of questions, you can use the accompanying rules to improve the act of asking a question.

Try these ten simple rules. Their use will help improve the clarity of your communication.

Ten Basic Rules for Asking Questions

1. Be direct.
2. Make eye contact if asking the question in person.
3. Use plain language.
4. Use simple sentence structure.
5. Be brief.
6. Maintain focus on the subject at hand.
7. Make certain the purpose of the question is clear.
8. The question must be appropriate for the situation and the person.
9. The manner of asking should reflect the intent.
10. Know what to do with the answer.

2. Asking Questions Is the Skill of Effective Management

Managers do not need answers to operate a successful business; they need questions. Answers can come from anyone, anytime, anywhere in the world thanks to the benefits of all the electronic communication tools at our disposal.

This has turned the real job of management into determining what it is the business needs to know, along with the *who/what/where/when* and *how* of learning it. To effectively solve problems, seize opportunities, and achieve objectives, questions need to be asked by managers—these are the people responsible for the operation of the enterprise as a whole.

All questions asked in a business setting are asked within the context of organizational expectations. This context is the "expectations of success" context in which all corporate discussions are conducted. I have yet to find a business that is seeking anything other than success, however it is they choose to define it.

This context of expectations defines the box that business people "think in." Success is specifically defined by the function, such as sales or research, or by the market, but the box frames the questions for each specific inquiry—or, in some cases, the inquisition.

Expectations of Success

QUESTIONS + ANSWERS = SUCCESS

The inquiry process is characterized as a linear model, and for our purposes in this discussion, it is linear. All the divergent thinking, outside-the-box thinking, or any other paths people may follow move along this general line—from questions to answers to results.

The process of asking a question within this context has eight basic elements:

1. What do we know?
2. What do we not know?
3. What are our objectives?
4. What do we need to know now to reach our objectives?
5. Who are we going to learn this from?
6. How are we going to learn it?
7. What are the expected results from deploying what is learned?
8. What do we do as a result of learning the answer?

This is the basic process along which questioning proceeds. Elegant models can be added to improve any aspect of inquiry a business might need. However, the focus remains the same, it remains simple, and it remains defined as success. If more success is desired, ask more questions. If the business wants to pursue a new business model, build a new box of expectations.

3. How Good Are Your Skills?

The numbers of brain workers, or nonproducers, as they are
called, should be as small as possible in proportion to the
numbers of workers, i.e., those who actually work....

—Fredrick Winslow Taylor,[2] father of modern management

"Brain workers" was the original concept of scientific management pio-
neered by Fredrick Winslow Taylor. His theories produced the foundation for
management portrayed as "modern" in the twentieth century. Think about
how much has changed over time. Brain workers are now the producers in
today's world of business.

Historically, managers possessed the knowledge, experience, and skills necessary to perform the tasks relevant to the daily operation of the business.
They could function as both bosses and employees. This competency was the
primary reason business owners promoted their employees to management.
The complex needs of the modern business have changed this model.

Such diverse knowledge is now required in business that an individual manager is rarely expected to be knowledgeable enough to run all aspects of the
business successfully without employee specialists. So, what do generalist
managers have to know to maintain the progress of their enterprise? They
must know how to ask questions.

How Good Are Your Questioning Skills?

While I was traveling around the world on a business assignment
that I discuss later in this section, I noticed that many managers
asked similar questions and got amazingly different results. The
way questions were asked appeared to be as important as the
question itself. I looked around for a book to serve as a good
training guide for myself on how to ask questions. The resources
I found fell into two categories: professional training guides (such
as for lawyers, teachers, and market researchers) and self-help
books designed to enable the individual to get ahead (such as
with interviewing skills, or improving a person's thinking
processes). These are all excellent resources. A number of them

are referenced later. However, my goal was to find a basic skills book. I was unable to find one that met my criteria.

When I started studying questions, I started with the assumption that I knew nothing about them. So, I built this book as a personal reference because I was unable to find what I needed.

After I embraced my own ignorance about questioning, I started to see questions in a new light. I found that even experienced, successful managers run into problems with their questions on occasion. They fall into traps such as habit questioning, posturing, or putting answers in their questions. Other managers, particularly new ones, commit a number of errors, such as asking prejudicial questions or leveling complex questions about interesting but unimportant or even unrelated details. Fixing these mistakes early in a person's career can lead to better personal performance over time. Fixing them among all managers can often lead to improved business performance.

The bottom line for all of us is that we need good questions because we want better answers. There was a need, at some point, at Enron, for example, for someone to ask the tough questions—*inquisitor's questions.* Investors needed someone to ask serious questions of the people at Global Crossing and at many other firms where damage occurred. It is not the job of any government agency to clean up these messes by asking business questions; that is the responsibility of management. *Management* is an inclusive word to mean anyone in a position of authority/responsibility—from line supervisors to board members.

Lives and careers have been ruined, not by questions, but by the lack of questions. As managers, we either do not know how to ask, what to ask, or are unable to ask the question for a variety of reasons. Sometimes we avoid certain questions because we believe that by asking them we risk our job, our status, personal embarrassment, or perhaps we are just being polite.

If managers at all levels were empowered by improved skills to ask questions sooner, better, and with an eye on what is best for the business or for their organization, disasters could be reduced and in some cases perhaps avoided altogether.

Management needs questions before it gets answers.

4. You Ask Too Many Questions

A downside to questioning should be explained: It is possible to ask too many questions, to ask them at the wrong time, or to completely misunderstand your situation; in many cases, the mere act of asking questions can even cost you your job.

·✍·

A young business division of a major corporation[3] was poised to introduce a new product, something the parent company had done, literally, a thousand times before. Only this time, a problem existed. Simply put, the product did not work. To make matters worse, no one appeared to be aware of this.

Was management blinded by the potential earnings from a product with a large gross margin? Perhaps it was the rush to get to market before the competition that caused such denial. Or was research, or manufacturing, or some other part of the company covering up the problem?

The situation came to a head about a week before the scheduled commercial launch.

Not a single person in management was aware of any problems. Everyone was focusing on the expected outcome—a boost in sales with a significant increase in earnings. The prospective new product had a large gross margin, and no competitive products were on the horizon. These kinds of opportunities do not come along every day, so all management eyes were on this product, this team, and the numbers.

The scene where the first inkling of a problem occurred was at the product team meeting. A half-dozen people were gathered in a small conference room at the divisional headquarters. They were conducting a rather perfunctory review of all aspects of the new product: technical development, manufacturing, marketing, sales, and service. The development process for this product had been flawless.

A massive report containing all relevant information sat in front of each person. The oversized notebook contained the marketing plan, manufacturing reports, technical service plans, global distribution plans, and a voluminous section filled with quality-control testing data.

Manufacturing was churning out inventory in anticipation of a worldwide release while the sales force was being trained on product benefits. Advertising and marketing materials were already distributed globally in a dozen different languages. All distribution plans had been checked and rechecked in preparation. This was all standard procedure in the company.

Everything had gone smoothly for the first-time product manager (PM) who was chairing this meeting. She eagerly anticipated the success of this product as a "career maker." It had all the attributes that PMs dream of: large market demand, lack of competitive products, high projected earnings, low costs, and an experienced support team to help get her over any rough spots that might occur. Up to this point, it was smooth sailing.

Most of her team were old hands; the manufacturing superintendent and the technical development manager were both 20-year veterans, and the quality-control manager was a professional quality engineer (QE) and had participated in dozens of product releases. A just-hired technical-support person was the final member of the team. He was to be responsible for managing the technical service effort that would support customers after they purchased the product. He had been on the job all of one week, after a month of orientation training.

According to company procedures, all members of the product team were required to sign the product release form before any new product could ship. Even the newly hired person would be required to sign—company policy mandated that he be considered a full member of the team, with the same responsibilities as all other team members in the review and release of product. After all, he was to manage the support of the product after it was released into customer hands, so his was a key role in the process.

Following a brief review of quality-control testing data, the new guy starts to ask some questions.

> **New guy** (interrupting the meeting to ask an obvious and somewhat foolish question): I noticed that all the numbers on the final testing chart are at the lowest possible limit for an acceptable release of product to customers. Am I reading this correctly?

> **Technical manager:** Where did you say you went to school? (followed by laughter)

> **QE** (mocking the new guy): Yes. So what?

> **New guy:** May I see the raw data from the testing lab?

> **Manufacturing superintendent** (pissed off by this young, inexperienced, ignorant new employee): We do not have time for this BS!

> **Technical manager** (acting highly insulted): Listen. As you can see, the data shows that the numbers are still all in the range of acceptable performance.

New guy (unaware that he is a major irritant to everyone in the room): That may be so, but the question I am really asking is have these numbers been rounded?

Quality-control manager (angered by the assertion that the data was somehow tainted): Yes, and all of it has been done correctly and according to proper scientific notation! You do have a degree, don't you? (more laughter)

New guy (undaunted by his obvious ignorance): How many product samples were tested?

Quality-control manager (annoyed and red-faced): Testing was performed on samples taken randomly from production inventory according to proper procedure. For college graduates, this means that the numbers are statistically relevant.

New guy (continuing to question the group, although he is by now conscious of the rising stress levels in the room caused by his questions): Although the data is scientifically correct, did any single sample of the product pass all five tests by more than the minimum?

Quality-control manager: Why you arrogant bastard!

Technical manager: Do you think we are stupid? There are well over 50 years of experience in this room, and what do you know after being on the job for 1 week?

New guy (now aware that he has a problem): The numbers we are using to pass the product for release represent a potential problem. They are all low. As a matter of fact, if the testing was as tight as the data suggests, then do you think that any of this product is any good?

PM (wanting to avoid a meltdown of her first product team): Time for a break. Let's all get some coffee and reconvene in about 10 minutes.

Collectively, the members of this product team were responsible for a consistent run of profitable new products without encountering any major problems in the market. They had complete confidence in each other and credibility with corporate officers.

The technical and manufacturing leaders escorted the new guy to the coffeepot. Many man-years and millions of dollars had been invested in this product. In addition, their personal reputations were at stake.

These senior managers were not about to allow this newly hired outsider to get in the way of their personal string of successful product releases. Their annual bonus and future compensation rested on these new products, which, at this point in their careers, represented a significant amount of money. They had seen similar situations before where products barely made it through testing but were acceptable in the market. The company was a market leader known for quality products. So, the test data came as no surprise to them.

Even if a problem existed, they knew it would be fixed as quickly as it was discovered. As a matter of fact, that was the role of the new guy. His job was to support this product once it reached the market. They did not take a liking to him.

The hallway discussion was over quickly.

> **Manufacturing superintendent:** You were assigned to this team to keep your mouth shut and learn. We do not think you are going to work out here. You ask too many questions. You should look for employment elsewhere.

And with that dismissal, the technical manager and the manufacturing superintendent walked back into the meeting room.

As the new guy was returning to his office, his boss (summoned by the PM, who wanted to avoid a nasty scene) met him in the hallway. Effective immediately, the new guy was put on a temporary assignment.

However, now uncertain about the results, the PM nervously reconvened the team without the "new guy." She was upset. This person whom she had never met before raised doubts in her mind that she could not dismiss as easily as her team members had sent the new guy packing. She elected to put off the final signing, on a technicality, to allow another customer technical-support person to be placed on the team. This proved to be a wise decision.

The product was never released. It did not work, just like the new guy had suggested after questioning the data. However, the business continued to persist in the development of this product because it was high on the wish list of their customers and the earning potential was greater than any other new product planned for the next two years. The invested dollars produced nothing of value. The product never went to market.

• (6) •

This is the story of how I started my career in industry. My reassignment, my punishment for saving the company from a serious problem, was to fly all over the world investigating and resolving customer product-quality complaints. I was to visit each and every site where customers had product-quality complaints that could not be verified in any way by normal procedures and resolve the problem.

I decided to accept the situation. How could I explain a painfully short tenure on my resumé? So, I was on the road (or in the sky) almost every day for nearly a year.

I was completely unequipped for this assignment. I was responsible for customers, and I was empowered to act as their agent. But, I thought this meant very little. I had no influence with anyone at the company. My immediate supervisor and the managers at the next level wanted me to quit and were doing everything possible to encourage me to find another job (so that they wouldn't have to fire me and run the risk of exposing their product problems). In spite of this, I discovered that I was able to have a significant impact on the company.

I found that I could help solve problems, reduce the number of customer complaints, improve products, and even increase revenue by asking questions. When I called back to the company on behalf of a customer, I questioned everybody and anybody about product quality. I had no problem calling the plant manager, the fascist superintendent who tried to intimidate me into quitting, or anyone else to solve problems for customers—my customers. I was prosecutor, inquisitor, and chief justice of the court of customer confidence.

After all, they had already fired me! I had little interest in the usual politics or politeness that normally gets people promoted out of such awful assignments. The only threat that the company held over me was to stop depositing money into my checking account. Although I knew that day would inevitably arrive, I was on a mission.

It was on this extended journey that I started to recognize the value of questions. Questions got actions. Questions also resulted in more questions. Questions caused people to think. Questions also made people uncomfortable, created stress, and could cause problems of their own.

I asked questions. I listened to the questions others asked in response to my questions. I listened to customers' questions.

Then, I started to keep notes on the "good" questions until I also realized that almost any question had the potential of having a positive or negative impact (often depending on who was asking and how). I also started to observe what managers said and how managers communicated when they asked questions.

The PM and I became reacquainted when I finally returned to the home office. She shared with me how upset she was by what happened in the project meeting. My questions had raised concerns in her mind that were just not there before I naively opened my mouth. Her instincts told her to be worried not only about the product, but also about the team she was running (and to be concerned about her career). She was on the fast track to be promoted before this bump in the road.

I had stumbled into her project meeting like some kind of drunk, spewing questions without concern for how my behavior might affect others. My questions may have been good ones—and it was clear that my insights were correct—but my approach left a lot to be desired. Our conversation went something like this:

> **PM:** They laughed a lot after you left the meeting. I felt like crying. After you raised doubts, I had to follow up, and you know the rest. You do know that my boss is the manufacturing superintendent's wife?

I was dumbstruck. It had not occurred to me that personal as well as professional relationships dot the landscape of businesses. My inexperience had contributed to a near-fatal career event for a talented businesswoman.

> **Me:** After I started to ask questions, to follow the trail, I just did not know how to stop. Also, I knew that even if I said nothing, I would not have signed that release authorization.

> **PM:** Well, at the time, I thought my career was over. But if we had let that product out, I would have been blamed for the whole mess anyway and might not have had a job to worry about. So, I did what a product manager is supposed to do—more teams and more meetings. I will get another project, but you had better watch your back.[4]

Her stressful experience, as a result of being on the other side of my questions, sensitized me to understand that I needed to fully appreciate the context, both personal and professional, as much as possible when asking questions.

Some managers were good at asking questions. Through insight, habit, or inquisitiveness, they made a positive impact on their businesses—on the people in their organizations. As I observed the management practices of both effective and ineffective managers, I started to break things down into bite-sized chunks for ease of use for me as a reference.

When I finally did reach management positions, I found this resource to be an invaluable guide. As I started to work more closely with leaders of one company, and then with leaders of other companies, I studied their questions, how they asked them, and what kinds of results they achieved.

This book is a distillation of those observations along with an analysis of questions as a management tool.

> If managers are looking for better answers, they must start by improving their questioning.
>
>

Common Errors: How to Recognize and Correct Them

5. What Are the Common Errors?

Have you considered what you do right and what you do wrong?

All people are not equally adept at asking questions. Many of us fall into a number of common traps when we ask questions, or we ignore the need to ask a crucial follow-up question. A trap is an error in questioning that we may succumb to without even noticing. As you will see in some examples later, others do notice. The most common of these so-called traps are

- Habit questions
- The question as an answer, or the answer in the question
- Asking a question without communicating the context

A number of other errors that creep into questioning might be the result of a conscious choice. We choose an approach to asking, we think of the words we will use, and we think about how to position the question in the discussion. But just thinking about them ahead of time may also incorporate errors. Some of these errors are intentional and might have nothing to do with asking questions, and everything to do with the process of asking. However, people are not blind to these practices, and although in some cases they make good theater, they might not help managers improve the business.

- **Positioning.** "I'm just a poor kid from Podunk, but tell me...."
- **Posturing.** Projecting an "I'm in charge" image.
- **Avoidance.** Ignoring the need to ask about the elephant in the room.
- **Casual questions.** There is no such thing as a casual question.

- Speaking "jargonese." Jargon is to be avoided.
- **The "no question" question.** A "leave them guessing" approach to asking.

Things can go wrong even when we do not ask questions. That is because we often conclude that it isn't the right time to ask. Or we conclude that the question and the possible answer are obvious to everyone, so we avoid looking stupid by not asking. If there is one thing that this discussion argues for, it is for asking questions whenever and wherever they seem appropriate. It is the only way to learn, to improve, and to foster a desire to do the same throughout an organization. At one time or another, at least one of these preconceived notions has occurred to all of us:

- You already know the answer, so it is unnecessary to ask.
- I will look foolish asking.
- Someone else will ask.
- I will wait for a more appropriate time.
- My question will make waves.

6. Do You Have Habit Questions?

This is the single most common trap that I have observed among all managers. I include myself in this characterization. Habits are difficult to recognize, and even harder to change. Experienced managers, particularly those who have had great success employing their favorite questions, are unlikely to change even when they are aware of the habit. To them, habit questions are not problems because the businesses they are responsible for managing continue to be successful. Consider this real-life scenario, for example.

Supervisor to his subordinates: We will be meeting with Jake tomorrow. Remember to have answers for his favorite questions: a) What does it cost? b) What is 10 percent of that number? And, c) if the budget were reduced by that number, what would not be done?

Jake had a one-track "10 percent solution" fixation. He asked his 10 percent question of every program and every person in every meeting he attended. He applied it across the board: advertising, manufacturing, research, and human resources. The business he ran definitely benefited from his habit.

However, he rarely asked any serious questions about other aspects of his business. His habits caused all internal discussions to be skewed in the direction of cost savings. In addition, funding requests were boosted by the same "10 percent solution" or even 20 percent increments because the cuts always came 10 percent at a time, and he always wanted to cut some cost.

How can you tell whether you have habit questions? Ask yourself, or better yet ask the people you work with. If people know what you are going to ask in most circumstances, you have a habit. This doesn't mean that it's a bad habit. It does mean that you can become more effective and your business can benefit by adding new questions to your toolkit.

I used to ask my kids the same question every day at dinner: What did you do in school today? One day, they erupted in anger. They were convinced that I didn't care what happened at school that day or any day for that matter. I often neglected to follow up on what they had told me the day before. We fixed the problem, and I am sure that on some days they wished I had forgotten to follow up.

Sure signs of habit questions include the following:

- You can write a list of your favorite questions without hesitation.
- You suffer from lack of follow-up on previous discussions.
- You are always impressed by how well prepared people are with answers.
- Someone has put a copy of this page in your mail.

If you have a habit of asking the same questions, you can avoid this by looking for continuity with previous discussions, or by asking questions from a new point of view. Read down the list of basic questions in the preceding section and select the one interrogative word you rarely, if ever, use.

7. Does Your Question Lack Context?

The context of a question is the general environment that provides both the questioner and the respondent an understanding of the expectations that come with the question and answer.

The best lack of context story I have heard comes from the White House a few administrations ago. The president's daughter (it is not important which president's daughter) had come home from school and asked for help on a homework question about South America. I can imagine how the story might have happened.

> **First Daughter:** Mom, I have this question about South America I need to answer for school. Do you know...?
>
> **First Mother:** Why are you asking me, dear? Your father is president of the United States of America. He ought to know the answer. Go ask your father.

Then, just like millions of other kids, she asked her dad. Her father did what all dads (who do not know the answers to fourth grade geography homework assignments) would like to do—he called the State Department. Tens of thousands of pages in reports were delivered the next day to the White House by truck.

Do you think the president had made it clear why he was interested in whatever it was he was asking?

If people are constantly asking for an explanation of the questions you ask, it could be that you are not providing the proper context.

Questions lack context when

- The respondent asks, "What do you mean by that?"
- The respondent is uncertain just how to answer.
- People always seem to misunderstand what was meant by the question.
- Your e-mail is full of information you never (thought you) requested.
- Someone restates your question by saying, "I think this is what Harry meant to ask...." (And your name is not Harry.)

The reason you asked your question should be as clear to the person you are addressing as it is to you.

8. Do You Put the Answer in the Question?

The question can sometimes offer the answer. Occasionally, this is the intent; but putting the answer in the question is recommended only if you are asking leading questions, and then only if you want to learn nothing more in the discussion. A variation of this approach to asking questions is called "putting words into the mouth" of your respondent.

There are a variety of forms of this practice, and the following example is an illustration of a manager's attempt to put the right answers into his questions and the appropriate words into the mouths of his staff. It happened in a business that produced and sold extremely sophisticated high-technology equipment.

The business manager had mistakenly promised the CEO, who in turn promised the board, that a new product would be released before the end of the year. This product was urgently needed to prop up sagging sales.

Overly optimistic business manager: We have maintained the product testing schedule, the customer beta testing, and no additional performance testing is included; so, we have acquired all the data we were originally planning to get? Right?

Product manager: Yes, we have all the data we were originally planning to obtain.

Overly optimistic business manager: Our suppliers have informed me that they are on schedule with all aspects of production, so raw materials inventories are at the necessary levels. Are we at proper supply levels to assemble the product?

Operations manager: Yes, inventories are at the planned levels.

Overly optimistic business manager: So, you all agree that we have all the major criteria satisfied? Right?

It is worth noting at this point that the business manager is asking questions that not only contain the answers but also questions to which only one answer can be given. He is not allowing any bit of undefined reality to creep in that could derail his plans for the promised product release to the market.

Overly optimistic business manager: So, the product is ready to be commercially released, yes?

Product manager (unmoved by this display of managerial brilliance): No.

Overly optimistic business manager: No? Did you mean no?

Product manager: Yes. I meant no.

No longer optimistic business manager: It's December. We need the product released to stores this fiscal year. We already have orders worth millions of dollars. I have promised the CEO, and he has promised the board. Everything is completed, and there is nothing standing in the way of sales. What happened?

Product manager (looking forward to a sales assignment in Baghdad): The final production prototype fell over, caught fire, and burned to a crisp. We don't know why this happened. So, until we finish root-cause investigation, everything is on hold.

So much for asking questions with answers, or even asking them in a way that you get only the answer you want. Reality may be different than desired.

In this particular case, the business manager did make one final attempt to influence his junior manager to start sales "*...anyway, just so we can say that we did it this year...*" and complete the necessary work during the first quarter.

However, the junior manager was unfazed by this ruse. He refused. To the credit of all, the product problem was resolved, the product was introduced successfully, and the managers were eventually rewarded with promotions.

Putting answers in questions, or words in the mouths of others, is best left to courtroom practice. Listen to the questions asked for these signs:

- Respondents always provide the exact answer you were looking for.
- The words in the question are repeated in the answer.
- The main question is followed by a second question: Right? Yes? No?

Unless a classroom lesson is being taught by repetition of a specific answer, it is usually a better practice to ask more open-ended questions. You might not hear what you want, but you are likely to hear what you need.

9. Positioning: "Just a Country Lawyer..."

Uttered in a casual manner, this preamble to a question is designed to be self-effacing, projecting the humility of the inquisitor. Instead, it may very well be viewed as disingenuous.

I knew a manager who used a variant of this. His preamble was a claim that he was a country kid from Tennessee. He used it so often that it became a signal that a "tough" question was coming next.

Manager: Help me out here. I am just a country kid from Tennessee. How is it that exchange rates are responsible for our decline in revenues in Japan, whether the dollar is up or down versus the yen? Can you explain this to me?

This kind of personal-positioning approach appears to have had its origins in the televised Watergate hearings. Sam Ervin was the primary investigating attorney during the televised congressional proceedings examining then-President Nixon's part in the criminal break-in at the Democratic Party head-quarters in the Watergate Hotel.

Mr. Ervin started many of his insightful, complex, and critical questions with his comfortable slight southern drawl declaring, "I'm just a country lawyer." What was the implication?

He wanted answers to his questions put in language so simple even a "poor country lawyer" could understand them. He was also positioning the other side as his opposites—slick, big city, fast-talking people.

Some of the sharpest minds I have ever met are possessed by country lawyers. Just like country doctors, they need to know everything about everything just to be prepared for anyone who may walk into their office.

Sam Ervin was neither poor nor was he from the country any longer. He was a Harvard-educated attorney and had earned a good living in Washington, D.C. However, his comments made great theater. He can still be heard speaking on a few different websites that carry recordings of the Watergate proceedings.[5]

The regular practice of positioning is discouraged. Here are a few signs of positioning:

- Everyone knows where the questioner is from.
- Employees can repeat the opening line of a manager's preamble, verbatim.
- A question is preceded with a remark positioning everyone else. (Okay, wise guys, let's get started.)
- The questioner uses us versus them, another positioning trick. (Why is it those night people always leave us the tough ones?)

The day shift versus night shift was used by one customer service operations manager I knew to improve the performance of his crew—the night people. He had heard his boss's boss extol the virtues of the day crew, even taking some credit himself for establishing higher customer-satisfaction numbers. "But you know how difficult it is to get the night crew working at that same level...." What an awful way to lower expectations while at the same time reducing the stature of the "night manager." It also said something about the attitude of the company toward customers who called at night.

The business was considering moving the night and weekend team to another geographic location to save money and, if necessary, to "enforce" better performance through the threat of lost jobs. No customer wants to hear poor language skills on the other end of the phone when he or she calls a service operation seeking support.

So, the night manager started to use a preamble whenever he addressed his crew: "Who is the day crew for the rest of the world?"

He used this whenever he started to ask questions, thus instilling within the team, through a simple question (one that needed no verbal answer), a sense of primacy in their work. To him, there was no day or night.[6]

A preamble has the most value the first time it is used. The direct approach is often the best. Ask straightforward questions in plain language.

10. Posturing: When the Questioner Suddenly Becomes Larger

Posturing is a sister technique to positioning. When a manager postures, he or she is attempting to create a larger, more important "self" than might otherwise have been perceived by the people present. Posturing might also be a way for an individual to assert control or gain influence.

A person "postures" by figuratively puffing up. This is done as if to show off a chest full of medals (like we would see on a heavily decorated military veteran). Posturing implies intimidation. It is one way to make everyone else in the room aware of who is in charge, or taking charge. Consider this scene.

Fifteen people were in a planning meeting run by a marketing manager. The discussion was designed to focus on market strategy. The objective of this get-together was to find a way to expand the business in a slow-growth market. All levels of the business were represented, from technician to a new vice president.

This was the first meeting that this particular VP had attended in this business and the very first time any of the people in the business had met him in person. A reputation for arrogance had preceded his arrival.

One internal wall of the large conference room was floor-to-ceiling glass. Everything in the room was completely visible to all employees in a large bullpen type of office area.

For purposes of our discussion, the business produces, markets, and sells hypoallergenic milk to schools.

Marketing manager (smiling): As you can see on our first slide, we have segmented the market into small, medium, and large schools.

VP (also smiling): Do you mean to say that these schools have low-volume, medium-volume, and large-volume milk-purchasing cafeterias?

Marketing manager: There is a direct correlation between the size of the school and the amount of milk the school purchases.

VP: Are there no exceptions?

Marketing manager (no longer smiling): There are some exceptions, of course. Some schools purchase milk as if they were larger or smaller.

VP: I see. So, is it possible to have a small school with a cafeteria that purchases milk at, let's say, a medium-volume level?

Marketing manager: Yes, I suppose it is. Now on this next slide, showing small schools....

VP (still smiling—grinning actually): Excuse me, but you said that the schools had low-, medium-, and high-volume milk consumption. You also said that some of these schools could have cafeterias that purchase higher or lower volumes of milk than would be indicated by the size of their school. Do these schools have signs out front stating "Large-volume milk-purchasing cafeteria inside?"

Marketing manager (with an expression akin to disbelief): No. That is not the point.

VP (grinning): That is the point. Why have you segmented the market by school size when you meant to do it by cafeteria volume?

Marketing manager: Because it just seemed to be logical based on what we know about the market. Now on the third slide, the needs of the medium school show that....

VP (the bastard was actually chortling at this point in the interaction): Why does this slide say medium school when you mean to say medium-volume cafeteria?

At this point, the marketing manager throws his pointer at the wall, makes a few choice remarks about some barnyard animals, how they should be segmented, and where in the market the new vice president can go to fetch them.

Much arm waving accompanied by hand signals of the marketing manager caught the attention of employees who were observing the deliberations from their cubicles just outside the glass wall. News of this performance filtered through the organization at nearly the speed of light, thanks to a new corporatewide fiber-optic network.

Through a Socratic question-and-answer technique, the vice president was able to accomplish many objectives: assert his control, demonstrate his self-perceived abilities of astute observation, and put a whole business on notice to get things right before they present anything to him. Finally, he was able to bring out the least attractive traits of an otherwise charismatic leader. This was pure posturing—and unnecessary.

The VP was smart, and he made certain to let people know that he was a member of Mensa, the organization of people smart enough to score over a certain number on a standardized IQ test. However, he was also considered poison for the business he had just joined. One person is not responsible for all performance of a business, whether good or bad, but it is hard to ignore the evidence in this case. Earnings and revenues started sinking, coincidently, shortly after the arrival of this executive and continued until he left the business.

I discuss the Socratic approach to management in a subsequent chapter. Although I am a proponent of the general method, it does have a dark side, as just noted.

Signs of posturing include the following:

- The question clearly puts the questioner in control of the discussion in a way that is unnecessary.
- Questions include qualitative measures of self-importance (for example, "When I was speaking to the president about this....")
- Attributes of self-importance are inserted into the discussion, such as SAT scores, Mensa membership, the size of their stock portfolio, who they had dinner with last night, and so on.
- The questioner explains the answers back to the respondents (for example, "Let me tell you what you mean....").
- Managers wear their Phi Beta Kappa keys every day on their foreheads.

Posturing is an unwarranted power trip that companies can ill afford. If it is necessary to demonstrate knowledge, reasoning ability, or a need to make corrections, ask questions that lead respondents to draw appropriate conclusions themselves. People respect managers who can bring out the best in them. Be aware, however, that although they respect them, they do not necessarily like them.

Managers must decide what is most important, but few people like a manager who is constantly posturing.

11. A "Casual" Question?

Managers can be the "manage by walking around" informal type, or they may have a more formal relationship with their organizations, or something in between. Questions, irrespective of the style, are always asked in the business context.

Whether a question is casual or not depends on how it is perceived by the person you ask. It is the respondent rather than the questioner who determines the nature of the question. Acting casual does not make the question casual.

The guidance I give anyone in business today is that there is no such thing as a casual question—especially questions put in writing or in e-mail. With the availability of e-mail records in the "forever file," any question you have ever asked, casual or not, is retained for some future trial lawyer to resurrect.

Questions such as "Ed, do you really think this stuff will kill all the fish in the ocean?" are not casual and will return to haunt you, your business, and perhaps even mankind.

If you are a person who manages by walking around, you might get away with casual questions, as long as they are part of your routine management style. If you are not prone to relaxed interactions or conversing casually with the troops, avoid engaging in this activity.

As I was staring into my computer screen one morning early in my tenure in marketing, our VP appeared in my doorway as if beamed there by a transporter. This was trouble—big trouble!

Here was a guy who left his office only to attend meetings for which he received a formal engraved invitation nine weeks ahead of time. He was ruthless when presented with a proposal of any kind—plans, advertising campaigns, or just plain memos. He poked, prodded, and otherwise opened holes in every paper, presentation, and argument presented to him. His stated objective was to peer inside you. When I saw him in the doorway, I went blank, and all external orifices slammed shut.

VP: Hey, I need an extra hundred thousand. As I understand it, all I have to do I come to see you, right? (He was attempting to project a casual, jocular manner.)

Me (after recovering from the shock of seeing him in the door): I don't understand.

He just stood in the doorway smiling. He would often do this in meetings: ask a question that was completely out of context and watch for the squirming to take place. Even when a person said he didn't understand, the VP would sit there and instruct them to "Think about it. I will just sit here and wait. I have all day." In this case, I just sat there, and he just stood there smiling and staring at me from my doorway.

Finally, after a long silence of about two years, I figured it out.

Me: Oh, do you mean the research funding committee I was just assigned to? (The job of this new committee, set up at the behest of the CEO, was to fund creative new technology development ideas that might not have gotten the attention of research management but that might still have some value for our markets.)

VP: Yes. I understand that our scientists can call you up and ask for money. Is that right?

Me: (Now I have come to the decision that he is visiting to banish me to our office in Namibia, which at that time was engulfed in a civil war.) Do you think the committee is a bad idea?

VP: I think that every person in my business should be thinking about my business all day long, all night long, on weekends, when they walk on the beach, and even when they take a dump.

Me: Do you want me to quit the committee? Just say the word and I will resign.

VP: I didn't say that. I just thought I would drop by to see how you were doing in your new assignment.

Me: I'm doing fine.

With that, he disappeared. This is an extreme example of noncasual questions in the guise of a casual interaction. The purpose of his visit was not subtle. He was reminding me that he was in charge no matter what assignment the CEO had made. His relaxed performance was unconvincing. Asking casual questions was beyond his personal limitations. No one in the entire business would ever accept any question from him as casual for two reasons: his style and his level.

Noncasual/casual questioning occurs when

- A business question is asked in an informal setting.
- An informal question is asked in a formal setting.
- Questions that "deliver a message" are offered in any setting as an offhanded or "by the way" type of remark.

Managers at any level are denied the opportunity, by virtue of their rank, of asking truly casual questions of their employees or of anyone at lower levels in the organization. Always view your question through the eyes of the respondent.

12. Do You Speak "Jargonese"?

All businesses develop shorthand communication techniques. They become local dialects—a language peculiar to that particular business or industry. As long as everyone knows the meaning of the words, phrases, or acronyms used, there are usually few problems with internal communications. However, jargon often creates misunderstanding and mistrust and can obfuscate meaning. Jargon may also be an impediment to better business.

Straight-talking companies outperform non-straight-talking companies.[7]

Deloitte-Touche, the well-known business consulting firm, has studied the use of jargon and reached the conclusion noted here. They even produced a software product called Bullfighter that reads documents and rates the amount of "bull" they contain. Although much of the focus of this scrutiny is on public documents, it is a good idea to drop the jargon internally, too.

Some words represent jargonese types of language that are employed by many businesses. Use of words such as *synergies, repurpose,* and *scoping strategies* lack clarity. Here is one of the oddest examples of jargonese that I could find.

Senior manager (asking about a business plan): Does that make you feel warm and fuzzy?

This is a direct quote. I am not kidding. To make this more astounding, people in his business appeared to know what was meant by *warm and fuzzy* when it came to financial reports. This expression became so popular it was used broadly across the entire organization. It was seriously used when strategic planning and financial forecasting was done.

How would one communicate with new employees, suppliers, customers, and government officials who would have no idea what *warm and fuzzy* means? The expression needs to be translated, and just how many interpretations of *warm and fuzzy* could there be?

The business was indeed warm and fuzzy, with respectable earnings for only one more year following the comment I heard. It was not the warm-and-fuzzy jargon that did in the business, but a hard-core cold and red bottom line. The business was eventually sold. The senior manager landed on his feet, became CEO of another smaller company, and both he and that company did well. I have no idea whether he continued to use his *warm-and-fuzzy* expression. I assume he did. However, the lesson of his previous tenure may have also taught him that no matter how warm and fuzzy you feel, you better not be seeing red on your bottom line or you could be feeling cold and blue.

Signs of jargonese include the following:

- Does it take time for new people to understand what you are saying?
- *Re* appears in front of words such as *purpose, strategy, resource,* and *think.*
- Simple words require long explanations.
- Statements sound like they have been written by politicians

The use of simple, direct language is the best policy. The objective of a question is to be understood so clearly by the listener that an equally clear response can and will be given.

13. Avoidance: If I Close My Eyes, Will the Elephant in the Room Disappear?

In many circumstances, the question that needs to be asked is avoided. Signs of this behavior have been seen all over the business community.

When an obvious question goes unasked, someone is shirking responsibility. When a manager, particularly a CEO, says, "I didn't know," in answer to a tough question about the business, it might be the truth, but it is a truth cloaked in deceit (by accident or intent).

A manager at any level who claims to be unaware of the deliberate actions of his or her organization is not asking enough questions. Or worse, he or she is avoiding the important ones.

In some cases, the manager actually places the elephant in the room. This is, of course, jargon—meaning that there is a large problem, so large, in fact, that everyone is aware of it, but no one is willing to say or do anything about it. Some companies call this a rhinoceros on the table—same difference. Large animal, poops where it wants; you get the idea.

I knew of a business executive who refused to allow any of his subordinates to present an earnings forecast that was one dollar below the preceding year's number. Not only would he not permit the presentation of lower numbers, he would not permit any questions that would even suggest that any current-year performance number would be lower than the preceding year.

> **Arrogant executive:** If you are here to tell me that you are unable to meet last year's numbers at a minimum, I will find someone else who can meet the numbers.

He meant what he said. No one ever mentioned the impossibility of meeting an earnings forecast because of his attitude. After a few years of success, he ran out of luck. The problem of earnings shortfall grew from an elephant to a woolly mammoth. The business failed in a way that no one could have anticipated.

His senior management team, without him, was snowed in at a small, homey New England hotel one stormy January evening. The warmth of the fire in the den, along with some excellent vintage wine from the cellar, relaxed the team so much so that they were willing to discuss the elephant problem: the lack of earnings. They held a meeting.

They knew that the effort to meet the preceding year's earnings had been successful. No one had discussed how this had happened. It just did, but with an unanticipated consequence.

> **Marketing VP:** How is it that we were able to make forecast? Did we sell that much more in December?
>
> **Sales VP:** I have no idea. We booked very few new customers at the end of the year. Most of our customers buy product under long-term contracts for better pricing. Where did the revenues come from?

Finance VP: I think I can answer that. We were all under orders to meet forecast—or else we would all be looking for new jobs. I saw that December was going to be short. So, I thought I could help.

Collectively: How?

Finance manager: We invoice customers when we ship. The bill is generated as soon as a shipment is placed on the loading dock, and the revenues from this bill are immediately included in the current month as sales.

Marketing manager: Exactly how much product did you put on the loading dock?

Finance manager: We calculated that it would take about 8 weeks worth of shipments, so we put out 12 weeks, just to be on the safe side.

Sales manager: We have loading docks that big?

This meant that the business was starting the new year with three months of revenues and earnings already accounted for with the preceding year's numbers. Although these people had worked as a team, they were forbidden, by executive order, to acknowledge the elephant—the shortfall in earnings. Yet they all knew that their substantial bonuses, as well as any raises and promotions, depended on meeting their forecasts.

You can guess what happened. Defeated by a management edict, the business went from tens of millions in earnings in one year to tens of millions of losses the next—all without a loss of a single market share point! A competitor eventually purchased the company.

Elephants are in the room when

- There are toxic topics that cannot be discussed.
- Some lines of questioning are forbidden.
- The business has a standing joke about what cannot be done.
- Winks and "you know" replace more verbal forms of conversation.
- The room starts to "smell" like an elephant.

> Questions should be asked when they appear necessary to ask. Although it might be unwise to take on the risk personally, if you see elephants in the room, it might be time to move.

14. No Question: Managing by "Wall"

Ever talk to the wall? It does not answer, acknowledge, or respond to anything you say to it. Some managers behave like walls. Police detectives are trained to talk to people who behave like walls, but most people in an organization do not have the luxury of such training. Behaving like a wall is a method employed by some managers to remain neutral. If you "manage by wall," find another method.

A Midwest agricultural products business had a manager who refused to make a comment about any presentation he listened to, any report or discussion he had—no questions either. He was fearful that any acknowledgment whatsoever would commit him to a position, pro or con. He would take a position on a subject only after divining the disposition of his management. The philosophy of corporate management at that particular time was similar to the old football philosophy of Ohio State's famous football coach, the late Woody Hayes.

Woody said that "three things can happen when the quarterback throws a pass, and two of them are bad."[8] Employing that philosophy, Hayes was one of the most successful coaches of all time. Although he did discover the passing game late in his career, his brand of football was affectionately referred to as "three yards and a cloud of dust."

The corporate version of that philosophy is "three things can happen if you make a decision, and all three are bad."[9] When this philosophy becomes standard practice for a manager, he or she becomes evasive. They may quibble with minor points in a discussion, "beat around the bush," or vacillate among positions without making any substantive commitments. Or, as in the case of the manager mentioned previously, remain unresponsive.

This type of equivocating behavior was adopted as the operative management style among lower-level managers of the firm. Decisions were made only at the top of the organization.

It is disheartening for people to work without feedback from their managers. Most of us want some kind of response during a discussion (and particularly when presenting information and recommendations). An acknowledgment is all that is required much of the time. Even if a manager values the attribute of inscrutability, it is wise under most circumstances to ask questions just to demonstrate interest.

One month after a friend of mine started a new job, he was asked to review his project for the boss. After days of preparation, he walked into a small meeting room where his boss was waiting—no one else. This seemed appropriate because he was going to discuss personnel, budgeting, and forecasts. The boss had a reputation of being difficult to read. He was generally unresponsive and rarely asked any question that would permit people to know what he was thinking, or whether he was thinking.

Three minutes into the 30-minute presentation, his boss fell asleep! Not knowing what to do, he continued.

As soon as he completed the presentation, his boss awoke, thanked him, and walked out. No questions, no comment. This, he learned from his peers, happened routinely. This was a "zero feedback" manager. You cannot get any more *zero* than this.

Some managers avoid asking questions because they do not want to tip their hand, or are uncertain how the question might be received. So, they don't ask. Asking a question is a sign that the time spent was of value. Without questions, there is little growth or improvement in people, processes, or business.

Ask questions, even if you are asking "just to be polite." It shows that you value the time of your colleagues—and lets them know you paid attention.

Neglected Questions

15. If I Ask a Foolish Question, I'll Look Foolish

If you do not ask the question, you will almost certainly be foolish—eventually. The only way to improve questioning skills is to use them. More often than not, others are also thinking about the same question and do not want to ask.

One of my former fellow managers was an ex-government administrator who had a tendency of allowing acronyms to slip into his conversation. Most of the time, they were easily discernible. On occasion, it was not possible to figure out what he meant.

I decided to look foolish in front of my boss and peers one afternoon when a particular acronym confused me.

> **Ex-military officer/manager:** Our plan called for us to lead a price increase—get the announcement out to our customers before our competitors had a chance to react. That is OTE.

I had never heard this expression being used before, so I assumed that everyone knew what this had meant. But, that did not stop me from asking. I had, as you will recall, already been fired once, so I was unconcerned about this happening again.

> **Foolish me:** What is OTE?
>
> **Boss:** Yes, what is OTE?
>
> **Ex-military officer/manager:** Overtaken by events.

I learned three things. First, my boss had no idea what this expression meant, and he did not want to appear foolish. Second, I learned that I did not feel foolish after I asked the question, only before. Finally, I picked up a neat expression to use as a label on many of the files in my office.

Foolish questions are often neglected and should be asked.

16. Unasked Questions: If You Already Know the Answer, It Is Unnecessary to Ask the Question

The reasons this behavior should be avoided are obvious, but not to everyone. The most common are these:

- You might be wrong.
- Others may need to hear the answer.
- It may be important to instill confidence in the person to whom you are questioning by "getting one (answer to a difficult question) right."
- Asking an obvious question may raise a nonobvious but vital answer.

The most comical performance by a manager I ever witnessed was of a person who would state the question and then, right after asking, explain that he already knew the answer.

When he also stated the answer, he was invariably wrong. Unfortunately for him and the people who worked for him, he refused to accept corrections. Even if his staff argued with him, he would simply state that he believed that he was correct. That was all that mattered to him.

In some situations, managers must recognize the need for others to draw the same conclusions or to learn for themselves what the managers themselves have already discovered. Asking a question for the purpose of helping an individual or a team to draw their own conclusions is sometimes a great way of exercising managerial responsibility to strengthen the organization by avoiding the tendency to act as the authority figure.

Junior people in some organizations may also need the chance to develop confidence by being put on the spot with a question. They need practice to develop the skill of thinking on their feet. Questioning people to allow them to deal directly with difficult questions is one way to assist in their career growth. Using this as a management technique might occasionally lead to an unexpected answer.

What a manager may think is obvious might not be obvious to others. The only way to find out whether this is the case is to ask the question, and then ask yourself why wasn't this clear to the others in the group.

An old and well-practiced habit of lawyers is to avoid asking questions of a witness unless the lawyer knows the answer. However, good cross-examination practices do not always translate into good business practices.

Managers do not have to know the answer to a question before asking it, and even if they do, it may still be worth asking.

17. Someone Else (of Higher Authority or Greater Experience) Will Ask

If you have a question, and it is clear in your mind that it needs an answer, ask! Unless you are a mind reader, it is impossible to know whether anyone else is thinking of the same question or will ever think of the same question.

Of concern to first-level and mid-level managers is their position in the hierarchy. Many questions they hear senior managers ask are good examples of the kinds of questions managers should consider. I have heard a number of mid-level managers express the opinion that they are in no position to ask tough questions.

They explain that their bosses often ask "those kinds" of questions due to their senior positions. The way to think about whether a "boss" kind of question should be asked is to consider this question: Is it important for the business to know the answer sooner or later?

If it is an answer that is needed sooner rather than later, you had better get it out of the way. Also, you need to keep in mind that the question might never be asked by the boss. And then what?

The credibility of the person asking is a secondary, but important issue. This is significant particularly in companies where deference is paid to people who have the battle scars of veterans, or "dirt under their fingernails" so to speak. Questions about sensitive subjects or a question that can produce a potentially embarrassing response should require younger or less-secure managers to pause.

Although important, is the question important enough to cause harm to yourself or to others? Or if your question will not be taken seriously in a public setting, such as a meeting, consider a private communication—by phone or in person. E-mail has a habit of being misinterpreted and should be avoided if possible under these circumstances.

People emerge into positions of higher responsibility, greater influence, or leadership over time. In many cases, their emergence is a direct result of the questions they ask. Answers are important, but the question plays a vital role. People who ask good questions are the people who learn to ask good questions.

One research scientist at a high-technology firm is often invited to sit in on business reviews, planning sessions, and many management discussions. He views most of these requests as a waste of his valuable time. However, the use of his time in this way is best for the business.

His perspective is so unique and his questions so insightful, he has become an integral part of the management process of that business in spite of the fact that he wants no part of managing. However, he asks stinging questions. So unconcerned about who he embarrasses and feeling impervious to any threats that may be directed his way, he focuses his laser thinking on issues he sees. The way he figures it, he cannot lose. The sooner he exposes the "fraud of the argument," as he likes to refer to management discussion, the sooner he can get back to the lab. He also does not care whether he is fired. His pension is already assured, based on his length of service, so he keeps it up. The net result is that it is always assumed that he will be at meetings, and this causes an appropriate amount of pre-meeting preparation.

Everyone wins in this setting. Management wins because the best information possible is likely to be presented. The presenters benefit because they will have done their homework, not wanting to be embarrassed in an open forum.

So in at least this one example, people have come to rely on "the other guy" to ask questions because in this organization, they have an official "other guy."

Someone else may or may not ask. If the business needs to know now, now is the time to ask. If the circumstances are not quite right, defer the question, but under no circumstances should you allow an important question to remain unasked.

18. Saved Questions: I Will Save My Question for Another More Appropriate Time

There is no more appropriate time to ask than when the question occurs to you. This is true even when you know the answer will be delayed, such as with voicemail. The rule of thumb that managers should adopt is to ask questions early and often.

But, the advice in the preceding section recommends putting a question off if the circumstances are not quite right. This is contradictory advice and something that all managers, when they arrive in any management position, quickly find out is all too common.

Yes, there is no time like the present—so ask now, while you still can. And yes, the question can also be put off, but not saved.

Saving a question implies that there is no real urgency for the use of that question. Questions are not like money. They do not grow interest in the question bank while awaiting withdrawal. Rather, their job is to produce a return on investment as soon as possible.

The previous recommendations suggest that a phone call or face-to-face meeting with the respondent is most desired, if possible. If not, e-mail or text messaging is the next best approach. This should be done as quickly as possible before events occur that might have been avoided had the importance of addressing the question been communicated. All businesses have inquisitors. Not all of them sit with business teams. Some are on boards or in research, but they are not always available to ask the tough questions so that others do not have to.

There is another exception to the "ask as soon as it occurs to you" rule: any question that might be asked in anger or with an intention to do harm. These should be put off—for good. What kinds of question are these?

We all know them well.

"What kind of an idiot are you?" is one of my favorites. I witnessed a senior executive explode one afternoon at a business review. He screamed this into the face of a mid-level manager. The manager had just finished explaining that he had authorized the construction of a $100 million manufacturing facility without doing any of the usual preliminary work to make sure the new

process would actually work. Normally, the business built a small-scale pilot plant to prove out a new manufacturing process. But, these small plants could cost up to $20 million, a significant cost penalty to a business with slim margins. However, to make a mistake bypassing the smaller facility could risk losing an investment of close to $200 million.

"A highly paid one," was the quick reply. This "wise guy," a middle manager who risked the company's money by building a facility using an untested process, was later promoted to vice president. The manufacturing facility never did open; the process didn't work. If only more questions had been asked by the senior VP, instead of venting anger and allowing humor to replace reason, the business might have avoided building a very large white elephant that was later sold off for about $20 million—to another company that wanted to use it as a pilot facility.

Questions might grow into problems when left unasked, even though asking them is no guarantee of being problem free.

19. My Question Will Make Waves and Making Waves Is Bad

Your question, depending on when it is asked, could indeed make waves. However, if it is an important question with a solid business purpose, it should be recognized as such and asked immediately. Of course, this is altruistic crap.

I would not ask anyone to sacrifice his or her career just to ask a question that can make waves. However, on occasion, a wave now is better than a tsunami later. Think about Global Crossing, Enron, and other companies that have ruined people, sent executives to jail, and hurt thousands of loyal hard-working employees.

Where were all the questions? Who on their boards was responsible for corporate governance? Why did abnormal numbers (whether unusually high or unusually low) go unchallenged? Perhaps someone attempted to challenge them and make waves. Perhaps they did not. We can never know the answers, and someone will eventually write the definitive business cases of these incidents. However, we do know that wave making is necessary in some situations.

Once again, the question that is neglected might be the one that saves the business. Of course, it could also be the one that gets you fired.

I was called into my boss's boss's office one afternoon. This was not unusual. Bill had a good rapport with the organization and had called me in on many previous occasions. This time, well, let's just say this time I was surprised.

A review of his favorite project had just been held in the morning, and I had challenged some of the underlying assumptions of the project. This was normal for me—I challenged the assumptions of the project I was working on, too. My job at that time was to develop markets for new technology products where customers were sensitive to any product deficiencies—the medical market.

So, my general habit was to quiz the product team on the details of the product deemed important to customers as a result of exhaustive market research. Although I worked primarily with my product team, I occasionally spent time with another major technology development team. They were used to my habits.

I sat through a review of the "other" project—the one that was "not mine" but was extremely important to the business. As a matter of fact, it was the number one product priority for the business. The product team I worked on was developing another "number one" most important product for the company.[10]

I sat in Bill's office, and he said, "I want to ask you something." He then drew a picture of a man standing behind a tree with a rifle showing a bullet whizzing out of the barrel. His sketching ability was quite impressive.

"Does this look familiar?"

"No."

"Is this you?" he asked, pointing to the sniper behind the tree.

I was speechless. I had no idea what he was getting at, although I had a feeling that it was *not* a good thing. His picture was pretty good. He may have missed his real calling.

"Are you sniping at the team?" Oh, now I got it. I was a little slow and just dumbfounded at his behavior.

"No, Bill. I do all my firing in full view—in public. I don't snipe." I got up and walked out of the bastard's office. I had already been fired once by this company, and it mattered not if it actually happened for real this time. The team was building a product that had some serious flaws, and they needed to be challenged. A product problem in the market would affect not only this particular product, but our whole product line. To me, this was a clear case of my losing the battle by being fired or transferred away to enable the business to win the war.

Two weeks later, another assignment was immediately made available to me, in a completely different division of the company. Plus, a promotion accompanied the transfer. I was happy to get away from the mess that was about to be created. The promotion was also welcome. However, this is not a recommended strategy for getting promoted. It is more likely to lead to your dismissal.

Bill's favorite product suffered cost overruns, delays, and was a market failure when it was released to the test market. The business was damaged by this and other failures in his division. He squelched any question that appeared to be in the way of his strategic plan and his favorite projects or people. Consequently, all critical questioning dried up. The parent company eventually sold off the business.

Rule of thumb for asking questions that you know ahead of time will make waves: Ask what is at stake? Then, look at the size wave that will be generated and what could possibly happen if the question goes unasked.

Then ask yourself, "If making waves can or will save the company, prevent a catastrophe, perhaps result in the saving of careers or pensions, is that bad?" It might be. In this case, even if I had been fired as a result of asking the question, it would have made no difference. Almost all the people working on the new development projects lost their jobs—it was just a matter of time.

· ⑥ ·

While traveling in France, I stopped in a small bakery for a croissant. The baker tried to interest me in some serious pastry. At that time, my travel and entertainment budget was beginning to make my suits shrink, so I politely declined.

"No, thank you," I said, "I am trying to be good," as I grabbed the portion of me that was starting to drip over my belt.

"Ah, what is good for one is bad for another," said the French baker.

So it is.

20. Normalization of a Defect[11]

This high-technology problem in questioning comes to us direct from NASA.

A problem or defect may be observed so often that it fails to generate a question. Because it is "normal" to expect the problem, questions go unasked. If a business can list "normal" problems, it has normalized the defect.

The origin of this expression is found in the official discussions about a space shuttle disaster. Foam insulation had broken off of the launch vehicles with nearly every launch of a shuttle. These foam pieces, now projectiles, hit the tile underbelly of the shuttle. This was viewed as a normal part of shuttle launches.

Managers, scientists, and engineers stopped questioning this defect. They expected it to occur and started to understand it as part of the launch process. Questions were raised, to be sure, but the questions lacked the kind of disciplined attention necessary to stop the problem from occurring.

When a disaster occurred, when a shuttle was destroyed with all hands lost, the agency was shaken out of complacency by the tragedy. There are instances of this in all businesses in one form or another.

I know of one not-for-profit institution that routinely loses checks. They are not really lost. It's just that they end up on the wrong desk, are placed in a wrong file, or sit in their envelope unopened—sometimes for years. This situation appears to be tolerated because the amounts are small and there is a steady stream of checks coming in every day. Most of them are indeed small, but in the aggregate, over time, the total is quite large.

Ask questions the first time and every time, even when there are no answers. Any "normal defect" should be challenged!

Misuses of Management Skills: Inquisitions Are Not the Only Abuse of Questioning

21. Errors and the Misuse of Management Skills

Errors can creep into questioning skills even if managers are successful in avoiding the traps, adopt a positive approach, and avoid making faulty assumptions. In this sense, management skills seem to be present but may be misused. These are the four common mistakes seen among managers with good questioning skills:

- **Assuming.** The assumptions made by the questioner may be incorrect. This is the most difficult of errors to avoid because it is difficult to explain the underlying assumptions of every question you might ask. The cure for this is to use plain language in the question and ask for the response in plain language in return.

- **Asking about details if they are not critical to the discussion.** They might not be remembered, or these questions may even impede the focus. Ask for details if they relate to the purpose of the discussion, the meeting, or report. If a detail is necessary for decision making, it's important to ask for it. If it is not critical and only of passing curiosity, meaning you have no intention of doing anything with it, avoid asking if possible. The question will connote an importance to the details that might not be desired.

- **Taking things for granted.** A fleeting thought may cross your mind and then it dissipates. You have taken something for granted. Ask the question.

- **Manager as the expert.** Avoid being the expert unless it is your job. For example, a small team of scientists, led by a manager who passes questions to the team rather than answer them directly, is much more effective than a team where the manager handles both the questions and the answers.

22. Is Your Question an Abuse of Power?

This is a variation of the corporate inquisition. Inquisitions are all about power, and this extends to situations that bear no resemblance to the meeting held by the trident-wielding inquisitor you met at the opening of this book. Abuses of power may be cloaked in the guise of looking for answers and can have results that are just as bad for the business as if the person were more clearly behaving like a devil.

Many managers ask questions that represent an abuse of the power of their positions. These are questions that would otherwise be unacceptable to ask among peers or would embarrass the manager if published on the first page of The Wall Street Journal. Consider the following situation.

A business team was presenting a new strategic plan to a senior manager, a leader in the company. This was a conservative company where numbers were checked and rechecked, and all optimism was downgraded to reflect an aversion to risk. This particular senior leader was an anomaly among her peers. Her behavior was more reminiscent of a gunslinger than the corporate banker mode of behavior shared by the rest of the senior officers. Her steep rise to the top echelon was beginning to be noticed by the more junior members of the organization, which gave her an air of invincibility, great influence, and power.

She was prone toward optimism and speculative behavior, and worked mostly from her gut with a "the hell with the numbers" attitude. Fortunately for the company, she had a habit of being correct and, better yet, delivering the forecasts she promised.

This is a reconstruction of the conversation. The names and identifying details have been changed to protect the guilty.

> **Senior manager:** What is the largest annual revenue projection you can make for the business?
>
> **Middle manager:** We forecast it to grow to be approximately $850 million.

Senior manager: Is that all? Don't you believe you can do any better than that? (bringing her fist down on the table for added emphasis)

Middle manager: That is a conservative estimate. We think there may be a considerable upside to well over a billion.

Senior manager: Oh, come on now. Can't you lie to me? How large can this be in your wildest dreams? (arms open, palms up, as if to show the size of a fish she just reeled in)

This is an improper question. Even if it is being asked in a lighthearted manner, it is a mistake many times over. Yet, this is a quote—it is exactly what a senior manager asked of a middle manager. She had asked him for a lie. She had asked for unfounded speculation that was totally out of character for the company and for the people who worked there.

One might argue that the success of this manager over time had educated her gut, so to speak, to be able to sniff out substantial growth opportunities when she saw them. Perhaps. However, her years of success might have been serendipity—luck. Moreover, just think about the impact an influential and highly regarded leader can have on all levels of an organization, on people who may follow her lead.

If you ever hear this kind of a question, run screaming from the room. This is not a manager you want to be standing near when the SEC[12] arrives. She will point to the person who lied to her. If, on the other hand, you have asked this question of others, you must enter a rehab program.[13]

You can and should ask people to imagine, guess, reach—all positive actions. However, if there is one rule for asking a question in any context or circumstance, it is to *never, ever ask for a lie!* The word implies all that is potentially bad in the commercial world, even if used in jest. Jokes are sometime taken as proxies for desired results, so the use of this word should be restricted to comedians.

The appropriate answer to the inappropriate question of "Can you lie to me?" should have been "No. I cannot lie to you. I am incapable of lying." This answer is, of course, a lie. Everyone is capable of lying, except perhaps Vulcans and Androids.[14] However, this kind of a lie is more defensible in a court of law.

So here's what happened.

Middle manager: We have numbers to suggest that, over time, this business could grow to between $1 and $2 billion, but a lot of things have to happen right.

Senior manager: I am very impatient. I don't want over time. I want to be excited about supporting your plan now. I'm still not excited about this. Make it a bigger lie. How large is this market right now—the total market, not just the targeted segment?

Middle manager (looking around at his team members, who are avoiding his desperate nonverbal screams for help): The market is about $9 billion in North America and $15 billion globally.

Senior manager: And you say this market is growing between 10 percent and 12 percent per year?

Middle manager: Yes.

Senior manager: And are you telling me that you are unable to see us leading this market on a global basis? (hands together in a prayerful pose)

Middle manager: Well, we could.

Senior manager: Okay. What is the largest number you can make up?

Middle manager (taking a final look around the room for help from his team members, who are by now in self-induced comas): $4 billion.

Middle manager: Now you have my interest! (She thrust her index finger in the air to emphasize the point, while appearing to skewer the middle manager at the same time.) This is exciting!

What is the danger in this? The senior manager could start to act as if this number were reliable. Worse, the team might try to make their business case fit this new number when they are clearly unprepared to do so. What happened in this case?

Well, the senior manager did just that. She began talking about this business as if it could actually reach the $4 billion mark, near term! The team, anxious to preserve their mid-level management positions, started to develop plans to support this hyperbole. They staffed, planned, and spent money as if $4 billion in revenue were to be delivered with the next shipment of office supplies. After draining millions of dollars in profit from the company, the business collapsed. The senior manager is now an executive with another company. The middle manager and many nonmanagers from this part of the business are also working at other firms.

Abuses of power do not have to be as obvious as this case. They can be subtle, casual, or even humorous. An abuse of power question is one that people are compelled to answer only because the person asking is at some higher level in the organizational structure. Period.

Check on your questions using simple criteria. Although you might feel as if you are not abusing the power of your office, the answers to these questions may give you a different perspective:

- How would you feel if this question were printed on the front page of the next edition of *The Wall Street Journal?*
- Would you want to be quoted asking the question?
- How would you react if the same question were asked of you, by your boss?
- Do you have to justify the question with a defensive preamble?
- Do you tell people not to take the question too seriously, but to answer anyway?

Questions are powerful—they are management's version of power tools. Anyone in a position of responsibility must treat them as such.

23. Are There Questions That Should Not Be Asked?

Yes, a number of questions should not be asked.

Some questions should be avoided under almost all but exceptional circumstances. They fall into the categories listed here (along with some examples). Most of these are common sense, such as the first category: Avoid questions that intentionally mistreat anyone.

Others are subtle, such as prejudicial questions—questions that offer a judgment as part of the question, and that judgment is unnecessary. Try to avoid these types of questions:

- **Questions that belittle, demean, humiliate, or otherwise cause harm to another person**

  Q: Everybody else can figure it out, what's wrong with you?

  Q: What is it about you and yellow paper? (Personally, I like yellow paper, and it's easier on my eyes. I once had a manager say this to me in front of the CEO. I felt stupid, and I switched to green paper instead. He hated green, too.)

- Questions that are personal

  Q: Are your kids always sick?

  Q: Tell me, Art, are those reports difficult to read, you being dyslexic and all? (This was a power abuser. He actually asked this question of one of our analysts.)

- Questions that add nothing or are unrelated

  Q: What do you think of my shoes?

  This is the management version of "do you think this makes me look fat?" Managers must not ask for personal or subjective judgments from their employees. It is one thing to ask whether a tie is blue or black if you are known to be color-blind and if you have an open relationship with your staff. It's another matter to ask whether the tie looks good.

  Q: How long do you think this full-day seminar will really run?

  Q: When will the market open tomorrow?

- Questions where a positive or negative answer may mean the same thing

  Q: Do you think that we should do this or not?

  Q: Is it, or is it not correct?

- Questions asked in the negative

  Q: Can't we get this done easily?

  Q: Isn't there any way this can be done?

  Q: What don't you like about (it, that, me, the dog, or whatever)?

  These kinds of questions provide insight but are asked in a way that positions the respondent as a whiner, and no one likes a whiner. Better to ask what was liked, to avoid focusing on the negatives. Does it really matter?

- Questions that can potentially backfire on you

  There are questions that can backfire on the person asking them. The rhetorical question, for example, might be answered in an embarrassing way.

  Q: You really wanted to lose the Smith deal, didn't you?

  A: Yes, I did actually. I really don't like Smith. He has monkey breath.

- **Prejudicial**

  Asking questions that are prejudicial can result in an adversarial confrontation. These types of questions create fear in people concerned about their future. They are harmful to individuals and organizations. Do managers really ask deliberately detrimental or hurtful questions? Unfortunately, yes. I heard the following question asked of a high-level manager by his general manager.

  Q: Why is it that all the incompetents ended up on your team?

- **Too many questions at once**

  How many is too many? One question a minute for an hour is certainly too many. Each situation will differ.

- **Constrictors**

  People can be influenced to "clam up" as soon as certain questions are asked. If a sarcastic tone is used at the same time, it can do serious damage to the relationships needed to smoothly operate the business.

  Q: How can you possibly suggest *that?*

  Q: Is *that* all?

  I once witnessed an executive vice president do this to a divisional president right in front of the division management team. It was an instant loss of credibility for both people: the VP because he demonstrated insensitivity and a hidden agenda; the divisional leader because she was obviously out of touch with her management, and this meant she was not likely to be there long. This was a shame because she had built a strong team. The business lost out in two ways on this one question. Employees no longer trusted the VP, and the talent and experience of the other executive left the business.

- **Ambiguous, misleading, and vague queries**

  How many times have questions needed to be rephrased? How many times have you seen someone try to answer the question he or she thought was asked, only to be interrupted by the questioner and corrected with the question that was intended to be asked?

  Q: Do these long-term trends mean anything?

  What the hell does this mean? This is the kind of question that usually requires a bit of narrowing down if the answer is to deliver any value.

- **Complex, nested questions**

  What is it with some people that they feel the need to ask nested questions—two questions in one? Most business interactions are not presidential press conferences where the questioner has only one opportunity to ask a question. Ask one question, and then ask the other.

  Q: When and where will the software package be introduced, and why is it not happening in the originally forecast timeframe?

  Q: When you stated that the protein was to be purified first, do we need it for targeting drug development, or is the process of manufacturing made more difficult when we are working with specialized solvents, and why wasn't this covered in your report?

- **Negative, multiple nested questions**

  A combination of two kinds of questions is to be avoided. It confuses the respondent at the very least. These kinds of questions are difficult to answer clearly.

  Q: It is correct, is it not, that your report shows your product line will not meet its objectives due to a delay of parts inventory because the product has developed problems for which you have no solutions?

- **Questions that impute negative attributes to the respondent**

  Q: Can you give me your honest opinion? (As if they would give you a dishonest opinion.)

  Q: Didn't you know that?

- **Defensive, qualified questions**

  Avoid qualifying your question with any type of a defense. It detracts from the question, and the obvious conclusion is that the question is offensive.

  Q: Excuse me for asking, but....

  Q: I hate to ask this, however....(If you hate to ask it, don't ask.)

  Q: This might offend you, but I must ask it anyway....(Find an inoffensive way.)

The rule of thumb is to ask yourself what the business value of the question is, and whether it meets the requirements cited earlier? Is it clear, simple, and so on?

Questioning: Improve Your Skills

24. What Are the Attributes of a Person Who Asks Good Questions?

In his book on cross-examination, Francis Wellman, one of the nineteenth century's most well-recognized attorneys, discussed the attributes of a good cross-examiner. Although observed more than one hundred years ago, these attributes reflect an ideal that we can apply to managers today. They share the status of "professional questioner" along with attorneys.

Wellman's attributes have been adapted specifically for managers of today and for anyone who relies on avenues of inquiry to excel at their job:

- **Attention to following up.** This attribute is vital for managers.
- **Ability to judge character.** This attribute enables the manager to choose effective questions that fit the respondents.
- **Ability to act with force.** The manager must know what to do with the answer and then have the ability and commitment to do it.
- **Instinct to discover weaknesses.** This attribute is an asset in all processes that require the use of questions.
- **Appreciation of motive.** Every business setting is accompanied by a set of motives for the business, management, and each participant.
- **Good intuition/instincts and business acumen.** This attribute is actually a set of attributes that combine to provide insight into what needs to be asked.

- **Clear perception.** This attribute enables the questioner to maintain focus on what is important.

- **Knowledge.** To a person skilled in asking questions, this attribute can mean knowing what is not known.

- **Ingenuity.** Business processes are unscripted interactions that require flexibility of mind and purpose.

- **Patience.** All minds are not equally facile, so appropriate time must often be allowed for discussions to develop to the point that questions are effective.

- **Logical thinking.** A thoughtful process is used for finding solutions to problems, which often means asking a series of questions and following through on details that are unexplained.

- **Self-control.** You must be able to exercise self-control even when others are losing theirs.

- **Caution.** Finally, caution in this sense is the ability to assess, understand, and respond to the risks of the circumstances.

25. Are You Prepared to Ask?

Whereas some managers prepare for meetings or interactions, many do not. Preparation, although not a requirement for good management, is advisable in most circumstances. Managers attend meetings, read reports, listen to messages, and, for the most part, react. They neither prepare questions nor do they prepare for questioning. Some of them will surely succeed because of their experience and skill.

Higher-level managers can usually use this strategy if they have a strong supporting cast of mid-level talent, so preparation might not always be required of all managers in every setting.

I have seen only two successful approaches to lack of preparation by a senior manager. One approach is followed by a leader of a health-care agency who enters each meeting he attends clearly uninformed about the details of the discussion. He allows the attendees to speak and will ask them to continue discussing the issue at hand until he formulates his questions and takes the lead. His years of experience in this agency provide him with the kinds of precedents he needs to be able to recognize patterns of problems and similarities in issues. He then uses this history to guide his organization through problem solving. This behavior is well known to all members of his staff. However, he keeps them from taking his lack of preparation for granted by occasionally preparing for a meeting at random. Because his staff members are never certain when this will happen, they can ill afford to neglect to prepare themselves with the required data. He is intolerant of poor preparation by others!

A second approach, from an educator, is more a question of the appearance rather than a true lack of preparation. This manager moves from subject to subject with ease, carrying only a small notebook that is never opened. She asks thoughtful questions and is often underestimated in that she does not refer to notes. However, she reads copious quantities of information each evening.

These approaches are not recommended for most managers. Both of these people are experienced managers who are prepared by years of experience on the one hand, and have an incredible appetite for information on the other. So even here, when there may appear to be a lack of preparation for a specific subject, the managers are prepared.

There are, however, certain occasions when reacting is literally all that can be done. These may include surprise employee complaints, unanticipated hostilities in negotiations, sudden changes in leadership, and any type of confrontation that pits a manager against a unique problem or situation. The character I mentioned at the opening of this book, Dr. Doom, was well practiced at informing his management of bad news at the last possible moment, in decision-making meetings. But in most instances, preparation can improve the quality of any interaction.

Here are some recommendations for managers to prepare for asking questions:

- What is the purpose of the paper, meeting, presentation, report, voicemail, e-mail, or communication?
- Why you are participating, reading, or listening—what is your role?
- Questions to prepare:
  Why are you specifically involved?
  How do others perceive your role?
  What results do we expect?
  What are the significant issues, as I know them?
  What do I hope to learn as a result?
- Are the important questions written down?

- What response is desired?
- What will you do when you get the answer (particularly what you might do with unanticipated answers)?

It is advisable you avoid reading any question if you are operating from notes. It is acceptable to refer to a list; doing so impresses people because they think preparation was done in advance. However, reading the question shows that you might be overly concerned with semantics—that you're more concerned with getting the right words than you are with getting the right question.

26. What's the Purpose of Your Question?

Asking questions, as simple an act as it may seem, can constitute a surprisingly subtle and effective management strategy.

—*John Baldoni*[15]

John Baldoni, in a monograph he developed for use at Harvard, outlines a number of reasons managers may ask questions. The list of possible reasons for asking a question is limited only by the number of questions that can be asked—meaning there is no limit. It is also more complex than just considering what you, the questioner, might have in mind. The listeners and respondents may read unintended inferences, implications, and management strategies into your questions.

"Why did you ask?" is the bottom line. It could be as complicated as "gut instinct" if you are probing for some hidden problem, or it could be a simple question about exchange rates. All specific questions are asked in the context of some general condition, situation, or event.

The general context of a question can fall into any of these business categories: planning, marketing, manufacturing, sales, service, technical support, human resources, finance, research, development, information, or other business-specific designator such as waste removal.

The business condition at the time of the question is a part of the general context, as is your position in the organization. We cover some additional issues about the role of a manager later.

A second element of the purpose is what it is you intended by the question. When you ask a question, you are addressing seven basic elements of purpose. Your brain covers these issues at nearly the speed of light:

1. **Data.** What do I need to know?
2. **Specificity.** Am I specific enough?
3. **Time.** Why do I need to know this now?
4. **People.** Is this the right person (people) to ask?
5. **Implication and inferences.** What are the possible ramifications of the question?
6. **Response.** What do I do in response to the answer?
7. **Manner.** How am I going to ask?

The mental process that answers these questions to satisfy your needs is fast, efficient, and often unconscious. Few people raise issues without having mentally gone through a similar list. One element is often ignored, and only after a question is blurted out does the manager think twice: inference.

Asking a question communicates a lot of information. Psychological analysis is not our intended purpose here. However, I have seen managers attempt to reel in questions that have gotten away from them. The manager might have wanted to infer something without stating it, and that's all right. Inference by accident is what you need to avoid.

Baldoni cites a number of general reasons for asking questions, such as "getting the lay of the land," planning, diagnosing problems, or making a change in mission. However, he raises two additional reasons that are worth mentioning: using questions to challenge the status quo, and using questions to encourage dissent.

Questions along these lines often go unasked. Decisions are often unchallenged, and elephants enter the room unnoticed. Consider asking questions to encourage dissent, or as a challenge when things are "going well" or when things are "going poorly"—two conditions when critical thinking can add great value.

You must ask them with an awareness of what is being inferred by asking. A manager who asks a plan or strategy be challenged must communicate, as part of the question, the purpose of the challenge. Otherwise, the manager runs the risk of actually undermining the implementation of that strategy.

The basic message is that a manager must always be aware of the purpose of asking. All questions from management, unless experience dictates otherwise, will be treated as if they have a distinct purpose, whether you have one or not.

27. Words: Are Some Words More Important Than Others?

The most critical need for attention to wording is to make sure that the particular issue which the questioner has in mind is the particular issue on which the respondent gives his answers.

—*Stanley L. Payne*

Stanley Payne was one of the founders of market research as we know it today. Wording, as Stanley Payne suggests, is important. You might think of this as one of those *duh!* statements. But, how many times have you heard a question asked, and then the questioner says, "Let me explain what I mean"? Why? Why should anyone have to listen to a question and then listen for the interpretation? Some managers make a habit of this.

They might use this "chance to explain" as a rhetorical device, a type of stall so that they can collect their thoughts while maintaining control of the discussion, or they might just be unprepared. The fact is that questions should be made clear enough when they are asked that interpretation is unnecessary. Word choice and use is helpful to consider in advance of interactions.

Some words, readily available for use, can go a long way toward helping managers improve their questions, and thus reduce the need for interpretation. Payne identifies in his market research work a number of good words to use in questions—words that can help lead to improved questioning and improved answers. For example, words and expressions such as the following:

Could...

Would...

Should...

Under what conditions...

And...

These words/expressions introduce an open question. They imply an interest in the information that is to be conveyed rather than in obtaining specific answers. They are good words to use early in any discussion process.

Other authors call for the use of "high-impact words."[16] These types of words provide an opportunity to give greater emphasis to a particular question than other words. Attorneys, for example, make use of impact words to help them establish meaning or to imply some intent as they build their arguments. Managers can also use them for effect rather than for building a case or impression before an audience. A few of these kinds of words appear in the following examples.

Q: What *exactly* is woffle dust?

Q: Is it a *guess* that our patent will be granted?

Q: Are you *certain* this material was from the same batch?

Implied standards, criteria, and performance measures can be suggested or communicated through a completely different set of words. It is important to consider the use for these kinds of words to give importance to the questions you ask.

Q: Are you *properly* monitoring the capital accounts?

Versus: Are you monitoring the capital accounts?

Q: Is this work *consistent* with standard industry practices?

Versus: Are you following standard industry practices?

Q: Are you making the *appropriate* changes?

Versus: Have you made the changes?

Q: Who supervised, and at *what time* was it completed?

Versus: Who supervised, and when was it completed?

If there are "smart" words, then you can be sure there are "dumb" ones, too. An article that appeared in the *Phoenix Business Journal* mentions "dumb words" used by salespeople. The words they cite are commonly used all over the business community, words such as *frankly* and *honestly*, which imply that the speaker might not be frank and honest all the time.

Examples of the Kinds Words
to Avoid in Questions

Frankly	Why would a person be other than frank?
Honestly	Are people expected to be honest in your business?
Best/worst	Needs definition. What do you mean by *best*?
Successful	Needs definition. What does *successful* mean?
Good/bad	Just what is to be learned from these descriptions?
Risky	There is risk in everything, even in doing nothing.

Open questions allow for a lot of latitude, but even then, words that need definition should be avoided. How does anyone know what the "best strategy" is? You know only choices, and even these choices face changing market conditions. There are no constants in business. Qualitative words offer an opportunity for respondents to equivocate.

> **Manager:** I thought you forecast successful distribution. How can you call it successful when all but one of our Ming Dynasty vases arrived broken?
>
> **Distribution manager:** Yes, it's true that only one arrived unbroken, but the vase was received on time by our most important customer.
>
> **Manager:** What exactly do you mean by *most important?*
>
> **Distribution manager:** The customer is the CEO's mother-in-law.

The words are important. Any time a manager uses undefined terms, or allows the use of nonspecific words in a response, it creates an opportunity for unmet expectations later.

An example of what not to do was found in a corporate training video for a well-known service business in which integrity is fundamental to the business. The trainer on the video actually promoted the use of these expressions for its customer service people: "Trust me...," "I have your best interests...," and "This is the truth...." These expressions are *poison.* Unsolicited

60

responses from people viewing this training video for the first time generated cynical comments from a small group of potential customers.

Try to avoid asking people to be frank with you or to express their honest opinion. This clearly communicates that you suspect that the opinions they have rendered in the past may have been dishonest.

All business inquiries should be earnest, using simple, straightforward language.

Simple words, used with emphasis or used in combination with other easily understood words, work best.

28. What Are the "Right" Questions?

What are my weaknesses? How can I balance them?
—Rudolph W. Giuliani[17]

If there were ever a list of the "right" questions for managers at all levels to ask of themselves, these would be at the top. These are questions that managers should continually ask. The answers can act as a guide for the manager on what to ask, of whom, and when.

I have already offered up a number of "wrong" questions. But, is the converse true? Are there "right" questions? The simple answer is, yes; technically, anything that is not wrong is right, but there is another way to think about this.

The right question is the one that yields the right answer, including its potential impact on others. We have already seen how asking a person to tell a lie can be a disaster, even if it doesn't have a negative impact on the person to whom it is addressed.

Right questions elicit right answers for a specific business purpose at the appropriate time and place in a manner that fosters the necessary rapport. A number of good resources are available that describe how to find and ask the right questions.[18]

Most of the time, the right question is a product of critical thinking, as suggested by M. Neil Browne and Stuart M. Keeley in their book, *Asking the Right Questions: A guide to critical thinking.*[19] However, "right" questioning is a situational skill. It is often the product of a collective thought process—the results of a series of interactions in a single place or over an extended period of time.

Understanding the situation is the key to understanding the right question or questions. There are always a number of right choices available, too—which might have a number of possible outcomes. It all depends on what you need to accomplish.

These guidelines may help in determining whether the right question has occurred to you or whether you might want to think more critically about the situation confronting the business:

1. **The question is meaningful.**

    It can be linked directly to a critical issue, strategy, or objective for the business.

    Q: How were we able to achieve a profitable year when, in fact, we were losing money at the beginning of December?

2. **It has impact.**

    Both the question and the answer have a potential impact, a benefit to the business.

    Q: With our line down, what do you think about contract manufacturing our product with our competitor down the street?

3. **You have made it clear why you are asking.**

    Q: I am asking to gain an understanding of how these things happen so that we can avoid repeating them in the future, not to engage in a witch hunt. So tell me, what went wrong and how do you think we can avoid this from happening again?

4. **The question matches the reality.**

    It is consistent with the reality of the situation. It is not out of character for the manager, the sense of urgency is appropriate, and it does not create skepticism in the person to whom it is directed.

    Q: Mary, we have known each other for years, and I have come to rely on your judgment in these situations. Is there any way we can save the deal?

5. **Know what to do with the response.**

    This applies to all questions under all circumstances.

    So why not trust your skills, your gut, or your experience? Why not just ask the questions that your gut instincts tell you are the right ones, or the best ones, or the questions you have seen work in similar situations before?

The danger is that our instincts may be wrong. Our instincts, by and large, are based upon our past experiences.

—Paul Schoemaker[20]

All business decisions will affect the future, and as every prospectus says about new opportunities, past performance is not a guarantee of future success. Actually, past success could conceivably cause future failures by blinding a company from asking or answering important questions.

29. Is Everything We Ask Important?

We would all like to think so. Reality is different. The way a question is treated, by our tone, facial expressions, body language, and gestures gives it importance. Yet, few questions are critical. The category a question falls into gives it importance as well as a perceived sense of urgency.

Three general categories of interest generate a business manager's questions:

- Things that affect the business in the short term
- Things that affect the business in the long term
- News

The first category of questions is the one most people respond well to. The second, although important, lacks urgency unless a manager gives it a sense of importance. Anything that does not clearly fall into the first two categories is a news item.

Managers who focus on the news elevate it to a status of importance that it might not deserve. It is a category of interest, and questions about the news should be treated as if they are just that—news. News is nice-to-know stuff, about entertainment, weather, or sports.

Q: Have you heard about the new robotic prostate surgery? Do you think it will affect demographics in the long run?

Q: What do you think about those floods in China? (when your business has no business in China)

Q: Did you know the boss is a Yankees fan?

Continued questioning with a sense of urgency concerning items of interest that lack ties to important business issues may have two effects. First, a manager who does this all the time is sending signals of an adult attention-deficit problem. Second, it might communicate that everything is important, ultimately having a negative impact when it comes time to focus on key concerns. Better to save the sense of urgency for those issues that require a high level of attention.

When asking questions, particularly of employees, consider which category the question represents, and then match the sense of urgency to the question.

30. The Manner of Asking a Question: Style

There is matter in manner.
—Francis Wellman[21]

There is a single unanimous recommendation from all sources on questioning: If you are asking in person, *speak clearly.* Include voicemail in this recommendation. too. Although you might not be physically present, your electronic residue is, and it represents you.

I've assembled a consensus list of recommendations here for consideration, borrowing heavily from texts designed for attorneys. Courtroom lawyers need to have the sharpest questioning skills; otherwise, their clients will suffer. Managers also need sharp skills; otherwise, their business will suffer.

These are basic and commonsense recommendations. Yet, due to habits or lack of attention, they are not always practiced. Consider, for example, this case of the CEO of a new business.

Whenever she was about to ask a member of her staff a question, any question, she folded her arms—literally, every single time. Her staff automatically braced themselves. She asked good questions, and they were highly skilled. However, the arm-folding habit always made them edgy and defensive. It took almost a year to break her of this habit.

The manner of the question, the way in which it is asked, and the actual communication of the question are all just as important as the substance of the question.

The Manner of Asking

1. Speak clearly.
2. Display confidence in your question.
3. Maintain good posture and pay careful attention to your own body language.
4. Avoid overstressing of certain words unless you mean to stress them.
5. Avoid superlatives unless you are leading to a conclusion.
6. Avoid exaggeration and hyperbole.
7. Use humor for a purpose, not just to be humorous.
8. Be brief; get straight to the point.
9. Know when to stop asking questions.
10. Listen carefully to the response. You may need to follow up or probe immediately.

31. What Was That You Said?

According to Stanly Payne, whose wise council guided market researchers for many years, a question changes depending on where the emphasis is placed.

Question	Meaning of the Emphasis
How *could* you say that?	Reprimand, as in "how dare you say such a thing."
How could *you* say that?	Other people might say it, but not you.
How could you *say* that?	You might think it, but saying it is another matter.
How could you say *that*?	Incredulity expressed over what it was you said.

The same question can have different meanings depending on where the emphasis is placed. I mention that point here so that you are consciously aware of it, and so that as you consider the phrasing of your next question, you can think about adding meaning by using emphasis rather than words.

I chose the "how could you say that?" example in this section deliberately. It was one of the standard habit questions of a manager I once worked with. He always asked the shortest possible questions and packed in as much meaning as possible. It was his way of challenging people even though he was a nonconfrontational manager. The technique worked most of the time.

Consider where the emphasis is placed when you ask a question. Can you add more meaning by using emphasis?

32. Can You Use a Raised Voice?

There are many right ways of asking a question. A military drill instructor may yell questions in the face of a raw recruit.

Drill instructor: Mister, what happened to your shoes? DID YOU SHINE THOSE SHOES WITH A BRICK?

Raw recruit: (responds by sweating profusely, while holding back a smirk because, after all, shining your shoes with a brick is a funny concept.)

Drill instructor: What are you laughing at, Mister?

A business manager should not yell in the face of a new employee—no matter how dull his or her shoes are. That said, sometimes a raised voice might be required. Some people recommend against raising a voice under any circumstances. *I am not one of those people.*

I believe it is permissible to raise your voice as long as you follow these simple rules.

1. Use a raised voice so infrequently that people will comment, "Wow. I never heard a raised voice before."
2. Avoid using the raised voice with groups, because it creates an "us versus them (you)" mentality (unless that is your objective).
3. The incident must be of sufficient gravity. Others who are within earshot must perceive you to be entitled to raise your voice.

4. Ask the question by speaking (yelling) directly at the person, looking him or her in the eye.

5. Do not hold back. If you are going to do it, do it like you mean it.

6. Try to use rhetorical questions. You are not really looking for answers when you yell, are you?

7. Avoid yelling contests. Deposit your rhetorical question and leave, without slamming the door.

8. Do not yell questions out of anger. Yell them out of purpose.

9. Maintain your self-control. Do not overdo it. Get it over with quickly.

10. Remove yourself quickly and allow the object of your ire to decompress.

Can I give you an example of when this manner of questioning might be used? Yes. What do you do, for example, when a person violates a direct mandate of the company, not once but three times? I can recall yelling only one time at the office, and it was because of this situation.

One of the more senior managers in my organization had thought better of a company decision even after we had a full discussion with our legal department. The decision was to end a business relationship with another company and to do it quickly and directly.

This smaller company had approached us to produce materials for a new type of construction product they intended to manufacture for outdoor use. Although we had technology that did work, and our research laboratories were able to turn out material that promised to be potentially useful, the business did not look economically attractive to us. The decision not to produce the material was made just as it was for many others.

A month later, I discovered the decision had yet to be implemented. This was not the raised-voice time. That happened six months later, after discovering it had still not happened. It wasn't a moment that I felt good about. Although I had been lied to repeatedly, I felt it was a failure of my management that it had occurred. I had assumed, improperly, that the matter had been resolved.

My questions went something like this.

Me: Steve, I never did hear the final response on your interaction with Universal Outdoor Flooring. What happened?

Steve: You haven't heard because I haven't told them.

Me: YOU WHAT?

Steve: They don't know yet.

Me: Are you going to call them now while I wait here in your office? Or do you want to come by my office in ten minutes and report on the call? OR DO YOU WANT TO BE FIRED? RIGHT NOW?

I gave him three alternatives and then walked out of his office, leaving him with alternative number two to do or number three to consider. Steve popped in my door ten minutes later—job done. I asked him to put a letter to legal, copying me, and send it with a return receipt to the other party. I no longer trusted him, and I told him so.

My judgment not to follow up after the first delay was poor, although I had a hard time believing than an experienced upper-level manager with more years in the business than I had would have done this. The underlying condition that caused the raised-voice incident was making assumptions and then not asking questions.

Although I support the possible use of a raised voice, according to the rules stated previously, it also behooves a manager to do whatever is possible to anticipate a situation such as this and prevent it from arising in the first place.

33. What Is Your Personal Style for Asking Questions?

Most managers have a general style when questioning people. It is their default mode. It's a habit. If a manager consistently uses one style, people start to rely on it. This is positive because it presents a consistent face for people to react to. Consistency is valuable in normal business settings. A consistent style, however, may be just as much of a problem for a manager as habit questioning.

It might preclude the ability of the manager to gain new perspectives or react to new circumstances. For example, if a manager always takes a neutral stance, others may be influenced to do the same. This is neither good nor bad—but people have a great tendency to emulate the successful managers that are promoted through the corporate system. Imagine a whole business full of neutral managers. How would they get anything done?

I used to watch a group of managers play a game called "Monkey." No one offered an opinion or would accept responsibility for any problem that was not directly under his or her control. The problem, or answer to a question asked by the boss that was about a business problem, became the monkey.

They would shuffle the monkey around the table so that it would land on anyone's back but their own. One member of the group would actually dance this virtual monkey around the table as if it were a marionette, prompting chuckles among the group—uproarious laughter if it actually landed on someone. Most often, the "monkey" was left forlornly alone, waiting, festering into a great ape of a problem.

This stalwart team was led by a manager whose style was 100 percent neutral. He was never flustered, nor did he ever appear to be swayed one way or another by any argument, no matter how persuasive it might be. He maintained neutrality because he was overly concerned about what the business leaders above his level thought. His management team adopted this style. "Monkey" was their game.

Problems started to pile up over the course of his two-year tenure. Lower-level managers and staff were constantly paraded in before the management team to present solutions, proposals, and projects for consideration. Everyone believed this guy was destined to become a corporate officer, and his style was being adopted by many people (to the detriment of revenue and earnings). The "Monkey" game ran full time until he received an offer from a competitor and left the company. I can only assume that he continued to be the Switzerland of managers.

Neutrality is, of course, only one style of asking questions and practicing management. The list in the following table is a general description of questioning styles as opposed to management styles. Some may be practiced as one in the same. The neutral questioner may indeed be the neutral manager just as much as the intimidator may be a style that works both ways, too. The objective is for each manager to know the style that is most comfortable for him or her and to then consider, when the circumstances are correct, adopting another approach by adopting an alternative style or styles.

Questioning Styles

Neutral	Controlled or relaxed demeanor, unresponsive to answers
Intimidator	Intense, put stress on respondent, body language is in your face
Investigator	Examiner, search for details, a "leave no stone unturned" attitude

Questioning Styles (continued)

Interviewer	Opinion taker, uses lots of open questions to draw people out
Interrogator	It is you versus whomever, extreme focus on the respondent
Commander	Asking as if launching or firing from a gun turret podium or throne
Grabbler	Asking as if rummaging around the closet
Quibbler	Argumentative through questions, edgy
Conductor	Directing the conversation through questions
Magician	Holder of the hidden agenda, respondent is guessing

Your style of asking is a combination of qualities represented by the labels in the preceding table. Rapport must be established with the respondent if you want the best answers. Each of these types of styles will generate different relationships. Even the intimidator develops a rapport with people. It just takes longer to get it established. One style has a distinctly negative influence and should be practiced with care: the magician.

Hidden agendas kill trust. If questions communicate a hidden purpose, or appear disingenuous in any way, the trust that is a complicit part of asking questions for managers will disappear. Magician questioners pull the proverbial rabbit question out of the hat to surprise their "witness." Use this with great caution, and if you find it necessary to employ, use it so sparingly that it is truly a surprise. A chief technology officer used to buy a different set of laboratory technicians coffee every morning. He used this opportunity to gather intelligence that he would use at key moments to ask challenging questions of his staff about yesterday's experiment that went awry. No one appreciated his magic act.

Answers will become guarded, and all of your skills as a questioner will be reduced in their effectiveness. If used consistently over time, you will be left with your rank and title as the only means to elicit answers.

To maintain their effectiveness, all styles should be based on these four basic qualities:

1. Be genuinely curious.
2. Practice your style actively by maintaining interest throughout the interaction.
3. Use patience—even a patient interrogator can get a lot of mileage out of her questions.
4. Project integrity.

34. Who Is Asking the Question?

What is your role in the organization? What are your responsibilities? The answers to these questions may give your question more or less weight than you intended.

Any question raised by a high-level manager in a business is quite naturally given a great deal of weight. The organizational rationale goes something like this: The manager must think this is important, because the question was asked. This kind of rationale often wastes money, wastes time, and raises more questions than it answers. Consider this example. The discussion takes place between a marketing manager (MM) and her subordinate, an experienced product manager (PM).

A vice president had torn an ad out of a magazine and sent it to the marketing manager with a question written in the corner.

MM: Have you seen this note from the VP?

PM: Yes. He wants to know if we are aware that there is a vacuum cleaner with the same name as our new product.

MM: What are you going to do about it?

PM: I wrote my answer right next to his question. See it up there in the corner of the paper?

MM: Are you crazy? One word?

PM: Look at his question. He wrote next to a picture of a vacuum cleaner that has the same name as our product: "Is this a problem?" My answer was "no."

MM: You cannot give a one-word answer—just saying no. He needs more than that.

PM: Why? He tore a page out of one of those in-flight magazines and wrote in the corner.

MM: That doesn't matter.

PM: You're right. It doesn't matter. That's why he gets a short response.

MM: We need to give him a comprehensive reply.

PM: Look, no one is going to confuse a tractor with a vacuum cleaner.

MM: But we have to make certain this isn't a problem. Have you?

PM: Yes. We have checked with everyone, received a legal opinion from our attorneys, and even called the vacuum cleaner folks to be certain. There is no problem.

MM: We need to explain this to him.

PM: I'm much too busy. It's not a problem. Don't make it one.

MM: Well, this isn't good enough. I'm going to check with legal and get a review of this right away.

Reassurance was what the VP was looking for. He said so, three weeks, thousands of dollars in additional studies, and many wasted man-years of meetings later. The marketing manager was reacting to the position rather than to the question. Was this an overreaction? Yes.

The simplest response would have been the best route to determining whether this was indeed what the VP was looking for. If more than a simple response was expected, he would have communicated differently.

The question and the rank of the person asking are always intertwined. They simply cannot be separated. So, ask questions that are important, and ask them in a way that communicates the kind of answer you expect.

In this case, a simple question on a torn-out page of a magazine should have been a clear signal that the VP was "just checking." He had confidence in his organization—that is how he made it to VP. No one gets to that level by himself or herself.

In this case, the one-word answer was all that the VP wanted, and is what he received from the product manager. What he learned from the marketing manager was that she might not be the right person for the position she was in.

Rank has its privileges, not questions. If they are clearly communicated, the answer should be expected to mirror the question.

35. Who Are You as a Manager?

Roles are not just a function of title. They may also be a function of the esteem accorded to you, your length of service, your special skills, or a host of other reasons. To get back to a point made earlier, there are no such things as casual questions in a business. They can be asked in a casual manner, but questions are perceived as a function of many aspects of the situation—the role of the manager being the key.

There is a hierarchy of roles ascribed to people that act as lenses through which their questions are focused on any respondent. Consider how you think about others you interact with, and then look in the following table at this list of roles played by people you interact with in a normal business setting. Any individual may hold a number of these positions.

Multiple Formal Roles of Questioners

Personal title	Mr., Ms., Mrs., Dr., Your Excellency, Prince, Duke, and so on
Business title	Manager, director, supervisor, chief executive officer
Position	Line management, middle manager, senior manager, leader, staff
Length of service	New hire, long term of service, deference based on experience
Personal relationships	Employees, colleagues, peers, friends, acquaintances
Situational position	Team member, advisor, confessor, adjustor, and so forth

It is important to remember that one of the most important aspects of the manager's job is to avoid taking things for granted. Many new managers can get off on the wrong footing with an organization by asking questions the wrong way because they might be unaware of their own "role set"—that set of roles attributed to them by others. The following approach to asking questions is not to be emulated. This was the approach taken by a newly assigned manager one month into her new assignment.

How do you do things around here?

Why do you do it that way?

Have you considered any other options?

On the surface, this appears to be a natural and innocent approach. The new manager was on an information-gathering mission. She was striving to learn all she could about the business she was now responsible for. The organization detected her separateness in the use of the word *you* whenever she asked a question.

She was treated with some deference because she held a doctorate degree and her business title was manager. However, she misinterpreted the deference paid to her as acceptance into her positional role of middle manager, her personal relationship roles with her direct reports, and the situational role she found herself in (trying to grow the business).

She persisted in using *you* when asking questions for the next few months. As a result, her team distanced itself from her. They offered only answers to her questions, and all without insight or explanation. The employees cut her no slack. They expected her to be a part of the team—as one of their leaders, and ask questions showing she was accepting the responsibility. It didn't happen. Her tenure was less than a year. What could she have done differently?

How does this work?

What do I need to know about this now that I'm here?

Who else might be able to help us?

How has this problem been resolved before?

Who has the most experience tackling this problem?

These questions have the same intent. It is equally important to avoid the use of the word *we*—at least at the beginning of a manager's tenure. In the case just described, things might have gone just as poorly had she asked, "How do *we* do this or that?" The use of *we* might be interpreted as sarcastic, or lack the genuineness of *we* from a longer-term manager. She failed in her new assignment, too. She thought her transfer was a reward, when, in fact, the upper managers were trying to determine whether this person was a senior management candidate. She didn't make the grade.

During the questioning process, a person takes on another role. Depending on the situation, once again, that role is somewhat different from the formal position one may be paid for. Probing, for example, usually shifts a manager into a different role automatically.

Informal questioning roles and their business functions are noted in the following table. Managers can move among these roles depending on what is called for in the situation confronting them. Assuming any one of these roles for just a moment can help establish new lines of questioning and important new perspectives.

Informal Situational Role of Questioner

Interviewer	To gain information
Teacher	To improve the knowledge of business
Student	To learn from the experience
Journalist	To gather information for the story of what happened
Detective	To gather pertinent information (facts, opinions, and so forth)
Prosecutor	To affix responsibility
Lawyer	To examine the facts, reasons, motives
Physician	To evaluate, find the cause, and prevent, treat, or cure the problem
Scientist	To hypothesize, test, analyze, and produce results
Historian	To uncover, to learn
Auditor	To examine and conclude

Two cautionary notes need to be sounded while we are discussing roles. First, the question should fit the person you are asking. Asking questions of someone who is incapable or not responsible for providing an answer is destructive. If you are looking for the cause of a problem and need details, for example, asking the hands-on employees, the knowledge workers, is appropriate. Asking for insight into strategic direction should be reserved for those people who have strategy development as part of their responsibilities.

The question should also fit the manager. Managers carry a responsibility with their role—from line supervision to senior executive. Impertinent, improper, or inappropriate questions detract from that role.

Signs and Signals

36. Hand Gestures and Other Physical Signals

Body language, hand gestures, and facial expressions are all signals. The signals you send are as important as the words you use.

Pointing, for example, is a key signal to people. Sean Hannity, a television news issues interviewer on Fox News, has what I like to refer to as an "attack finger."[22] During one interview, the person actually moved away from him every time Hannity pointed during questioning. The gentleman reacted as if he thought the finger was loaded.

Hands play a major role; open palms up, palms down, fist pounding, prayerful hands, folded hands—just be aware that your hand signals are consistent with whatever question is being asked.

This is not a text on how to translate body language—we all signal in some way with our hands every day of our lives. The issue when considering a question is the possibility of sending mixed signals with your hands when conducting an "interrogation." Here are five ways in which managers misuse hand signals when asking questions:

- Crossing your arms when asking open questions.
- Asking for an open-ended question like "Tell me the whole story" while making brackets out of your hands as if to contain the response within some virtual barrier.

- Hands held as if saying your nightly prayers, while asking about facts. One manager I knew would always hold his hands in that manner during financial discussions. I was never certain if he was praying for the numbers to somehow be "healed" or if he was just giving thanks that they weren't any worse.
- Hand waving.
- Sweeping motions of your hair or hand in the direction of the person you are addressing or flicking lint off of your trousers could be considered a dismissal of whatever that person is saying.

Hand waving has become a kind of physical business slang for "bullshitting"—and from my experience, these terms are interchangeable. Hand waving just sounds polite. The more signaling required during the asking of a question, the more obtuse the question is, and the more likely it is to encourage a confused response. Hand waving does not add meaning, because most managers are amateur hand wavers. If you want to learn from professionals, watch them. I recommend watching presidential press conferences, when the press is shown, to learn a few of the basics and how hand signals are used when asking tough questions.

There was a commercial on television in which a young employee is ignored after he makes a recommendation in a meeting. He is astonished when his words are repeated verbatim by a more senior person who delivers them with an aggressive hand motion—signaling resoluteness. When the guy claims he said it first, everyone is quick to point out that he did not say the same thing. "You didn't say it like this," another person said as he used the same hand motion. Although this pokes fun by using something so blatant that it is an amusement, it does have an element of truth.

I was invited to sit in on a meeting with two executives and a one of the best-known business consultants in the world. Problems had arisen between the two organizations these men managed, and the consultant, who happened to be in town, offered to spend an hour with them to determine whether he could be of any assistance.

He asked each to speak for five minutes before he would say anything. I was to sit quietly, saying nothing in spite of the fact that I had been invited by both to work with them on problem identification. After each had spoken, he announced that he had the answer. He was going to go buy me a cup of coffee while these two guys talked to each other for the next 45 minutes. With that, he stood up and motioned me to follow him out of the room. He closed the door on their protests that he was to help them.

Business consultant: Where can we get some coffee or tea?

Me: Right down the hall. I missed something. What just happened?

Business consultant: Did you see what Bob was doing while John was speaking?

Me: What do you mean, what he was doing?

Business consultant: Did you watch his hands?

Me: Yes, he was brushing something off his trousers.

Business consultant: He was dismissing everything John was saying. While John was speaking, Bob was picking the lint out of his trouser cuffs and brushing it off his pant leg.

Me: Now that you mention it, he did that the entire time John spoke. And John sat with his arms crossed while Bob was talking.

Business consultant: The body always signals what the mind is unwilling to say out loud. They don't need you and they don't need me in that room. They need to talk to each other. Their problem is that they talk past each other and don't hear a word the other says.

We returned 45 minutes later. Bob and John were having a discussion. It might not have been the friendliest conversation, but they were talking to each other. They had to; there was no one else in the room.

The hand signals by each one of these executives disclosed their inner thoughts without the need for a single hostile word. As a matter of fact, the discussions between the two had always been pleasant.

If you have an interest in the study of body language, a few sources are listed in "References" at the back of this book (refer to Martel, Finlayson, or Haydock and Sonsteng). A few basic rules can be applied to help your non-verbal questioning hand-signal language that can help your verbal questioning skills:

- Check your hand and arm positions to see whether they match your words. Open arms and open hands for open questions.

- If you are uncertain whether a hand motion is appropriate or what hand motion is appropriate, keep your hands still.

- Avoid pointing, gesturing, or motioning to one person in a group setting even if there is only one person who can answer your question. If you want to hear from a specific person, use the person's name. It engages the person in the discussion.

37. Eye Contact

There is general agreement among all sources that I have checked that eye contact is an extremely good habit to acquire when asking a question. Maintaining eye contact with your respondent is important (selected target, the person being quizzed, victim, or other) when you ask, as well as when the respondent answers. Do not stare, but look at the respondent. If you are asking a probing question, and it is a contentious discussion, you definitely want to look the other person in the eye.

I once had to confront an employee who we discovered was getting kickbacks from some of our vendors. He tried to engage in a staring contest, as if this would make the data go away (or perhaps to intimidate me into believing whatever story he was prepared to serve up). Instead of asking him the direct, closed questions I had planned and he had obviously prepared for, I changed tactics after asking one question. I asked open-ended questions.

Our discovery of his misappropriation was quite by accident. His mother had suddenly and unexpectedly become deathly ill, and he had run out of town in the middle of the night to be by her side when she passed away. That morning, a substitute secretary signed for an express delivery package. Thinking it was important, she opened it and delivered the contents to me. Inside were invoices and account summaries for his review before submission to the company. It appeared as if he was receiving a "commission" based on volume purchasing by the company from this particular vendor.

Auditing was notified and, as heartless as it may have appeared, we had a meeting his first day back.

> **Me:** Joe, why do these vendors show checks made out to you?
>
> **Joe:** I have no idea what you are talking about. Show me. (This terse response was accompanied by an extremely hostile stare, complete with clenched teeth.)
>
> **Me** (now looking at him): Joe, tell me about your relationship with these guys.
>
> (He was speechless. He had brought with him a number of files and papers, but was unprepared for an open question.)
>
> **Me** (now looking right at him again, right in the eyes): Joe, what were the circumstances that led up to this situation?

He opened up. I listened to a description of billing errors along with an explanation of how he was assisting these vendors outside of the workplace.

His discussion was incoherent. He was also looking away the entire time. He did not admit to any unethical or illegal behavior. Joe left the office in the middle of our chat and returned the next morning to resign.

This was an extreme eye-contact situation—uncomfortable for both parties. The point is that even under the most stressful situations, adhering to the basic rules of questioning is likely to produce the desired results.

Rules of Thumb

1. Eye contact is not staring. Avoid staring contests.

2. Contact at the beginning and particularly at the end of your question is important.

3. Look at the person to whom your question is directed. It makes no sense to look elsewhere even if something more interesting is going on.

38. Demeanor, Body Language, and Facial Expressions

A CEO of a small company actually used to take a step backward every time he was about to ask a question. Do you think the people in his company provided him with open answers? Do you think they worried about his reasons for asking?

If you want open, act open. If you want interest, lean forward and act interested. If you want firmness, use hand gestures to show you are firm (not intimidating). In one sense, the answers you get mirror the questions you ask—and the way in which you ask them.

This is not an exposition on body language, nor is it a discourse on appropriate facial expressions. If you have passed through high school, you have reached a point in your socialization where you can tell the difference between an expression of anger and one of amusement. What I want to emphasize here are basic rules for asking that will help you meet with all situations where you will need to ask questions. You can always add to your ability to "act as you want."

General Body Language Rules

1. Maintain open (uncrossed) arms.

2. Face your respondent directly when questioning.

3. Look a person in the eye when you ask questions.

4. Stand or sit erect.

5. Keep both feet on the ground.

6. No jiggling, shaking, or rocking. Agitation or impatience is communicated as much by body language as by expressions. If you want good answers, do not communicate agitation even if you are agitated.

7. Lean slightly forward immediately after asking. This connotes interest in the answer.

8. Keep a relaxed face for the best response. Avoid furrowing your brow, pursing your lips, biting your lips, squinting, wincing, frowning, or otherwise communicating discomfort.

9. Smile or look friendly. This works in your favor even when you may be probing for serious mistakes or unethical or illegal behavior.

10. Breathe normally. Heavy sighs right after you ask might unsettle the person.

11. Stay alert.

12. Appear prepared for the answer. Always expect the unexpected. Surprises happen infrequently, but they do happen. Prepare to take them in stride.

Questions are asked by your whole body. Answers will account for verbal as well as the nonverbal parts of your interrogation. Be sure to manage your whole interaction so that you are communicating in a consistent manner and not just going through the motions.

Types of Questions

There are four basic types of general questions: direct, indirect, open, and closed. An additional twenty different specific types of questions are variations of these four basic types. They may be asked in such a way as to make them fit under any one of the four basic types.

The kinds of questions that will become part of the normal repertoire of any individual manager will vary, depending on the role and position that a person holds in an organization. The most effective managers I have seen have always been able to meet difficult situations with a new question—one that the management team is not used to hearing. In other words, they always seem to be able to summon a different type of question when it is needed.

In my experience, most of the questions used in business are direct and closed—a reflection that "answers" are the primary preoccupation of management. There is nothing wrong with this as long as consideration is given to asking open questions (for example provocative, hypothetical, and divergent questions) when the appropriate opportunity arises.

For example, examining market opportunities is a time for more open-ended questioning, whereas closed questions are probably useful when evaluating root-cause analysis in a product failure. However, there are no rules. Many different types of questions can be creatively applied to situations to yield new insights, ideas, and action plans.

The purpose of this introductory discussion about questions is to provide a creative approach to considering what and how questions are asked. Even though there are a few basic types, the manager ought not to be limited by thinking that a limit exists.

Think of this list as a starting point for growing more questions.

39. Direct Questions

Most questions should be direct. Why? It is the easiest form of question to understand. A direct question is more likely to yield a direct answer. Managers generally have little patience for respondents who are long-winded, unclear, or indirect when answering a question, so why ask questions that suffer those same flaws?

Direct Questions

- Are easy to understand
- Are clear in meaning, intent, and purpose
- Communicate that a direct answer is desired
- Enable answers
- Show attention and interest
- Represent more focused control
- Are potentially stifling to introverted participants
- Put pressure on respondents

Q: What are the qualities of a direct question?

Q: John, how did that squirrel get into the server?

Q: Mary, do we have a plan for commercializing Rev B?

Q: Yes or no... now that we have heard the proposal, do you believe that a new ad campaign will yield the boost in sales we are looking for?

Asking a person a direct question by his or her name, especially in a meeting or public setting, is a positive practice. Notice during any presidential press conference that the chief executive often tries to do this when he calls on people, even though he is the one answering the questions. When you are in control of a meeting or when you are participating as a member of a management team, you have a share of the control by virtue of your position. By calling on people by name, you are exercising the positive attributes of control.

Calling on a person by name is a form of recognition. If the question is direct (easy to understand, clear in what you are asking, and so forth), the person you call on will feel good about answering. Of course, in some situations, direct questions make people feel uncomfortable, too, but the purpose of the direct question is always to get a direct answer.

40. Indirect Questions

In some situations, a manager wants to avoid sounding too onerous. For example, a problem might have occurred where it is less important to fix blame than it is to fix the problem. Alternatively, in other settings, there might be a need to pursue a relaxed or less-intense approach to getting answers because of the personalities of the people involved. In these cases, the indirect type of question can prove helpful.

Indirect Questions

- Generally used to establish rapport with the respondent(s)
- Allows a soft approach to controlling the discussion
- Indicates that the interest is in the answer, the "what"
- Permits others to "wonder," too
- May lead to new ideas or new lines of inquiry

Q: I wonder what caused the roof to collapse?

Q: Does anyone have an idea about how we can approach the question of gerbil migration?

Q: Do we have a concept for determining the frequency of this situation?

> **Q:** I really have no way of knowing how large the market is for woffle dust. (This is a question in the form of a statement, but it is an indirect question.)
>
> **Q:** Is there any way for me to resolve my uncertainty about marketing blue toilet seats?
>
> **Comment:** This sounds a lot like the Milligan situation.

The "I wonder" mode of questioning is a good one to consider. It is indirect in that it doesn't require a specific answer, and it is also open, allowing for the maximum degree of freedom in response. It was good enough to drive a Nobel laureate physicist[23] to discover the behavior of matter, it ought to be good enough to help us in the more mundane aspects of driving a business.

A statement rather than a question appears last on the preceding list. Many statements are just questions in disguise, as is this one. Responses may vary from head nodding in assent or disagreement, to people responding with comments. Either way, a response is what a manager is looking for. If none is offered, a direct question can always be asked. This is as indirect as a manager can get.

These questions all stay on topic but avoid defining the nature of the answer.

41. Open Questions

If you want to maximize the opportunity for any type of response, a direct open question is the method of choice. They are generally received positively and communicate a willingness to listen to whatever is answered in return.

Open questions allow for broad dialogue and free discussion. They are designed to draw out the maximum in response.

Open questions provide the best method for avoiding surprises in business. I have yet to meet a manager who really liked surprises, either good or bad. They all want to learn, in real time if possible, what is happening. The relatively neutral forum created by the use of open questions can allow employees to raise issues that might be circulating just below the surface of managerial awareness:

Q: Where do you believe the next competitive threats will come from?

Q: How can we learn about our customers purchasing preferences more quickly?

Q: "Well, Watson, what do you make of this?"[24]

Open questions are not designed to yield "yes" or "no" answers. They are asked specifically to avoid yes or no, and are usually asked to provide an unrestricted general direction.

Open Questions

- Usually viewed as positive

- Foster an expansive and inclusive participation

- Can cover a lot of ground

- Enable the telling of a whole story so that important details are less likely to be missed

- Challenge control and focus, which are more difficult to maintain particularly when time is tight

- May lead to grandstanding[25] by one of the more outspoken participants

Q: Tell me more.

Q: Please explain.

Q: What are other possible explanations?

Q: Can you describe the situation in more detail?

Q: Tell me the whole story.

Q: How does the process work?

Q: Please describe the campaign.

42. Closed Questions

Closed questions are *direct question* tools. They are generally used along with other types of questions. An entire discussion of closed questions can get tedious, unless it is a session devoted to examination of facts or conducted to probe specific events. Closed questions are used to ask for specifics: facts, opinions, details, or descriptions.

Q: What time was it when you arrived at the office this morning?

This question is completely closed. Although the respondent might know that you will ask about tardiness, or extol the virtues for arriving early, there is really only one option. You answer the question. The manager may or may not have a follow-up, and it might be a different follow-up question than the respondent is thinking about. That is why it is recommended that closed questions be answered as if under cross-examination.

Closed Questions

- Maintain focus
- Offer few options for digression
- Are used in searches, examinations, detail discussions, fact-finding, anything requiring specificity
- Are used to stop equivocating answers
- Narrow the discussion
- If overused, could result in "getting into the weeds" (too much detail)
- Include a lot of *what, who, when, where* questions

Q: What equation did you use to project that curve for the filtration rate?

Q: When, on what date, will this be completed?

Q: Who is going to be responsible for the report?

Q: What is your answer? Yes or no?

The following question types discussed in this chapter are just variations of the four basic types of questions (open, closed, direct, indirect).

43. Stupid Questions

There is no such thing as a stupid question. People might do stupid things, or have stupid ideas (which lead them to do doing stupid things). *Questions, by their very nature, are expressions of ignorance, not stupidity.* Asking expresses an interest in learning. Even when a question is asked as a way of teaching a respondent to think or provide an answer that a teacher, for example, may already know, the question is educating the respondent and providing the instructor with information about that individual.

One often quoted expression that the "stupid question is the one that goes unasked" is popular, but it's not the question that is stupid. So, any time you hear someone express the sentiment that the unasked question is stupid, don't be taken in.

A person can be acting stupid by not asking a question that cures ignorance. All of us are likely to have some personal experience with unasked questions. In the product meeting described in "Introduction: Questioning Is the Skill of Management," although I asked "too many questions," there were many questions I did not ask in this first meeting of a product development team that I had ever attended. The ones I did ask, although good questions, might not have been the wisest choice. They exposed the foibles of a group of senior managers who had been so successful that they had started to take success for granted.

From my current vantage point, I can tell that I was both ignorant and stupid. My questions were direct and closed. Had I considered asking other types of questions, or taking a different approach, the company might have avoided wasting more time and money. That particular meeting would very likely have ended the same way—with no decision. However, the members of the team might have been willing to listen to my questions.

That is all speculation and unimportant except as an object lesson. What is important for all managers to consider is how to avoid reflexive questioning and apply more thought to the situation.

The person might be ignorant of the answer, but the question is always smart.

44. Filtering Questions

A question asked specifically to exclude extraneous information is a filtering question. It is a type of closed question most often employed as a follow-up question or when probing for information.

Filtering Questions

- Are asked to exclude information
- Are used in probes and as general follow-up in less-stressful environments
- Are focused
- Begin to alert everyone to what is important and what is unimportant without having to spend a lot of time in explanation
- May result in challenges if others believe information being excluded is germane to the discussion
- May need an additional follow-up question or comment to provide clarity with what to do with the information excluded

Q: Which of these data points are the most important for us to pay attention to?

Q: How many customer complaints will need management follow-up?

Q: Which machines worked well today?

Q: If we ignore the data on this portion of the grid, what conclusions can be drawn?

Q: Which of these proposed product names should we cross off the list?

45. Double-Direct Questions[26]

The double-direct type of question is an exception to many of the rules noted earlier. It is a type of compound question, which puts words into the mouth of the respondent while leading down the path of inquiry chosen by the questioner. However, if done correctly, it is clear, easy to understand, and should yield good answers.

Double-Direct Questions

- Offer a way to maintain continuity of the discussion

- Can act as leading questions during a probe

- Can also be used to subtly probe for agreement, consensus, or for unmentioned facts, by enticing the respondent to disagree with the premise of the question

- Put words into the mouth of the respondent (which might limit discovery of information or squelch opinion)

Q: When all the genetically altered sharks got hives, what did you do?

The technique in this question is to use information from a previous answer as the basis of the question. However, it could also be that the person only implied that the sharks actually had hives, so the questioner might be putting words into the mouth of the respondent.

Q: Let me see whether I understand your conclusion. All the power went out while the production line was running, including the emergency power, and because of this, you were unable to determine if the fail-safe emergency stop was working?

This might appear more complex than it needs to be, and that might very well be what is wanted. The respondent in this case might already be defensive, so a question that forces a more thoughtful response may slow down the dialogue enough to identify key important issues that emerge.

Hypothetical Questions
(If, What If, Suppose)

The use of hypothetical questioning is a great technique for expanding beyond the limitations of a discussion. You can use these questions for a number of tasks: test strategies, consider alternatives, disagree without disagreeing, allow a minority opinion, and so on. (However, you must take care not to act on the hypothetical answers.)

Remember this advice when using the hypothetical. If, in one of the early examples, the senior manager had asked for a hypothetical answer on the potential size of the new business opportunity, she probably would have gotten the same magnitude of response. However, given her behavior (she had asked for a lie), it is unlikely that she would have acted any differently. She would probably still have treated the hypothetical answer as a potentially factual one.

Hypothetical Questions

- Expand the discussion
- Can be used to spark creativity
- Create opportunities for change
- Demonstrate openness to new ideas
- Are a way to think through scenarios
- Enable you to test data suggested in discussion without appearing unsupportive

Q: What if it were possible for us to double our market share?

Q: If the product had actually worked, what would our sales have been?

Q: Supposing cows could fly, what impact would that have on our umbrella sales?

Q: What if we were to license this invention to our competitors. What would that really do to our market share?

Considering a change in strategy? Try introducing the idea with a hypothetical question rather than a direct assault on the issue.

47. Provocative Questions

Use provocative questions when you are probing for information that the respondent may or may not part with willingly. Provocative questions do just that: They provoke. They issue a challenge to the respondent. The intended respondent either meets it, attempts to ignore it, or attempts to redirect it. A good provocative question will not be ignored, however.

These questions can be open or closed, but most will be direct. It is difficult to be provocative in an indirect fashion.

Provocative questions are also used to expand the thinking of a group. They can also be used to attack a problem, stimulate a moribund staff, and challenge the current management if you are a member of the corporate board of directors. Questions that provoke are too often thought of in negative terms.

They need not be argumentative or negative in nature. We often think of *provoking* as a negative attribute. In the case of a manager attempting to shake up an organization, get attention focused on the problem in a hurry, or just to engage everyone in the conversation, a question can be used to *provoke* or stimulate creative thinking.

Provocative Questions

- Present a challenge to the respondent or group
- Stimulate thoughts or new ideas
- Target specific critical issues in a way that gets attention
- Could encourage an emotional response

Q: We can fund only one of your projects, which one?

Q: Why should we believe the data?

Q: Give me one good reason why we should not sell this business?

Q: How can we double the output at half the cost?

Q: Why do we have thousands of dollars in unauthorized expenditures in your area?

Q: Have you ever had to lay off a large group of people before? (Although managers might not like surprises, questions are often used to deliver them when warranted.)

48. Rhetorical Questions

Rhetorical questions are asked for effect. Answers are not wanted. These are questions that contain the answer—by implication. The problem is that in many instances, this kind of a question can backfire and result in an answer the manager might be unprepared for.

Rhetorical Questions

- May be used for humor

- May be used as an accusation

- Vent anger

- Release frustration or emotion

- Are a deliberating device to share thinking with others without asking for an answer directly or indirectly

- Can offer theater, such as to be sarcastic

Q: Are you always looking for trouble?

Q: Do you think I should grow some hair just to pull it out with an order this low?

Q: Do you want me to fire you? Now?

> **Q:** Why me? Why do these things always happen to me?
>
> **Q:** What would this organization do without me?
>
> **Q:** What were they thinking when they signed that deal?
>
> **Q:** How could they do this?

A rhetorical question can, and often does, backfire when confronted with a wise-ass. "Are you always looking for trouble?" can lead a respondent to answer "yes." The manager in this situation now has to confront an employee who may be looking for a conflict, or this person may be the corporate version of the class clown.

Before asking a rhetorical question, it is always a good idea to consider how a person might react.

49. Reflective Questions

Reflective questions do exactly what the name indicates: They ask the respondent to reflect—to look back on a decision, on an event, a change in direction, or on information. The implication is that there is a lesson to be learned by doing this.

Reflective Questions

- Can return the respondent to a previous point in the discussion, or in time, or in action

- May teach a lesson without being didactic

- Link together information that provides insight into the current situation

- May be used as a way to pause the conversation

A pause elicited from a reflective question may have the effect of relaxing people by moving backward in time, when there might have been no vested interests in the room, such as the following strategy question. It may also have the reverse effect, actually intensifying pressure, such as in the "last time the plant blew up" question.

Q: Why do you think Ajax responded to our last price increase that way?

Q: Would we have been better off holding back?

Q: What happened the last time the plant blew up?

Q: How would we treat the problem now with this new technology?

Q: If we look back in our history, what stands out as the key strategic direction of the company?

These are not complex questions. They are generally open questions asking for "tell me" types of follow-up questions and story responses. The real value in asking reflective questions is to find the key to resolving a current dilemma.

50. Leading Questions

A discussion of leading questions is included under strategies. A leading question, although it can be deployed by itself, is usually part of a strategy—a plan for delivering a specific message or finding a way to get people to focus on the issue of importance.

When used by attorneys or by journalists—Do you still beat your wife?—it is designed to be more of a trick question, to entrap or to generate a response that provides a reporting opportunity.

Leading questions are really not recommended. Although they find their way into business situations in many instances, their use may be viewed negatively.

Leading Questions

- Are used in an attempt to generate the answer you want to hear

- Are often used to obtain agreement, even from recalcitrant people

- Enable a conclusion to be drawn for a group that might not want to come to a conclusion any time soon

- Drive decisions

- Can be used to reduce debate

- May push alternative opinions or notions out of consideration

- May be viewed by the respondent and others present as disingenuous

Q: The president's suit is a wonderful color brown, isn't it?

Q: The average customer waiting time on our hotline is too long. Do you think this is contributing to our loss of business?

Q: So, Mr. Burr, don't you agree that as you faced Hamilton with the sun at your back, his eyes must have been blinded by the bright sun?

51. The Pause as a Question

One of the most effective tools for eliciting more information—the pause—can be used in place of, or as part of, a question. When a comedian delivers a punch line, it is often set up by a pause. Setting up a question can work in much the same way.

It alerts all listeners to something different (and different is often interpreted as important) coming. The pause provides a natural break between a remark and a follow-up. It also serves to reinforce whatever the opening remark was because it requires the listener to make a note of it so that it is remembered in the context of whatever is coming next. If you think this is too subtle, just try using the pause and see what kind of responses you get, either from the pause itself or the question that follows.

Pauses

- Should be used after a statement that asks a question and encourages someone to jump into the conversation

- Can be accompanied by raising the eyebrows, a facial expression that invites comment

- Are occasionally greeted by silence

Q: So, let me see if I understand your recommendation. We provide the financing, and the first thing we expect to see is (pause)

Q: When the software was shipped, it was still in the shrink-wrap, but then it must have (pause)

Q: When we first observed that the pigs were, in fact, eating like pigs (pause) and we expected....

One of my fellow managers had the uncanny ability of raising one eyebrow so that it almost appeared to be a question mark framed against his nearly bald head. He would start a question such as "Is it a question of resource availability (raised eyebrow)...." This act generally precipitated a deluge of information. After his staff got wise to the technique, he then started to use the pause as an exclamation point, and then follow up the rest of the question, raising his eyebrow at the end.

   Q: Was it a question of resources availability (pause) or, was the system overwhelmed because of inadequate planning?

Although it appears to be a bit of a theatrical technique, managers are often called on to "play a role," whether large or small, in a meeting.

52. Silent Questions

A pause can act as a question during a statement, but the silent question is deafening in the way it can deliver a message. A research manager in a high-tech company used this method of asking her engineers questions all the time.

When data was presented that she was uncomfortable with, or just plain made no sense, she would lean forward and shrug her shoulders, opening her hands up as if to say "What?" However, she spoke no words. She would then wait for a reply. Although she used this practice quite a bit, it never failed to deliver—as long as she was in her element. The engineers had adapted to her style by responding just as if she had verbally asked.

When she employed this technique when speaking with the marketing organization, she tended to get silent replies. She then switched to the more common form of communicating verbally—asking a question to get a response.

A silent question, although an effective tool in many circumstances, does not provide any hint of what you need for decision making or what thought processes you might be focusing on.

Silent Questions

- Can be used as long as the people you are interacting with allow you to
- Are a good tool for one-on-one discussions where you want additional information without asking too many questions
- Enable you to mask your focus, which could be on something less important
- Enable you to avoid shutting off other avenues of discussion when the group might be swayed by inference

Q: Shoulder shrug, hands open raised slightly

Q: Frown, raised eyebrow accompanied by a quizzical look or look of puzzlement

Q: Open hand motion, as if to signal for more information

The important point here is that the manager is empowered by the other person when using a silent question or gesture as an interrogative. This is an interesting dynamic in that, as in the preceding example, the marketing folks just sat and looked at her and waited her out. It could be that they had seen her in action before, or it could be that they really just wanted her to commit. In any case, it's a technique best used in situations considered friendly.

53. One-Word Questions

One-word questions are not used often enough. People feel compelled to expound on why they are asking or attach a preamble to their question. A one-word question might be a bit theatrical, but the situation may call for it. Shocking news that has you in disbelief might yield a "Really?" or a "What?"

The side benefit of one-word questions is that they are open—as open as can be. They do not limit the kind of response and, therefore, allow the greatest possible latitude in reply. Their use also does not disclose any preconceived notion of the interrogator manager.

One-Word Questions

- *Why, when, where,* and *how* are the primary one-word questions.
- It is important to follow the one word with silence; otherwise, it loses the effect.
- The accompanying use of appropriate body language, facial expressions, and hand waving also add to the impact a one-word question can have.

Q: Really?

Q: What?

Q: Why? (a great question when a person renders a decision or offers an emphatic opinion)

Q: When?

Q: Where?

Q: How?

Not every setting is a good one for single-word questions. They can also become habit forming if effectively used by a senior manager. In a mid-sized company, a taciturn former operations manager had risen to take over the company. In doing so, he brought along his habit of asking one-word questions whenever he could. After a while, the questions wore very thin on his staff. Although he added the appropriate number of words, he took refuge in single words whenever he could.

The impact it had was to force people into presentation formats that enabled the questions of their leader, rather than the leader recognizing the kind of queries needed by management. Performance of this company remained mediocre.

54. Clarifying Questions

A clarifying question is another type of a closed question whose purpose is to make information understandable. Some managers use clarifying questions to refine the message they are hearing. Others use them as a way of building a case for a difference of opinion. Senior managers and leaders of companies or organizations often use clarifying questions as a tool to influence people rather than order their organizations to take specific actions.

Think of a clarifying question as a "Robert's Rules of Order" type of query. Legislative bodies use these rules to control the process of interaction common to the houses of lawmakers. A hierarchy of these kinds of questions establishes the concept that one type of query takes precedence over another in a debate or ordered discussion. Although effective, this process is not conducive to management.

One thing that you should be aware of and avoid is the use of clarifying questions to manipulate the comments so that they align with what you want to hear. A division director of an electronics business continually manipulated his staff with questions to "clarify what was meant by…," causing distrust to grow among the talented managers of his group.

He was overly concerned with any message in any presentation that did not perfectly align with what he had already explained or told to senior management. He would ask clarifying question after clarifying question. "Tell me exactly what this means." Or "Could it have meant…?" And then he would go on to describe what he wanted the response to be.

Clarifying Questions

- Are appropriate when the data is unclear
  or when certain terms are unclear
  or when an opinion is desired and the discussion does not indicate which way an individual might be thinking

- Can be used to identify a bias that might be swaying the information presented

Q: Can you clarify what you mean by *perfect stirring?*

Q: What do you mean by *we?*

Q: Exactly what are these commonly accepted accounting practices you keep referring to?

Q: By IRS, you mean (you might wish to use a pause to clarify)...?

Q: What are the critical issues I forgot about?

His personal strategy worked very well. He did become an executive officer in a company—a company that has done very poorly since his arrival. (The performance of the firm might be completely unrelated this individual, but his habit of clarifying everything he hears remains.)

55. Divergent Questions

Divergent questions are used to expand the number of possibilities without changing the subject. These questions usually change the perspective of participants or the direction of the discussion.

Every discussion in a business has a number of different constituencies. Instead of thinking like a business if a discussion is stuck on a particularly knotty issue, some creative managers often find it interesting to ask what other constituencies might have to say about the same issue?

This may include government, beer distributors, public policy groups, military, stockholders, and so on. One manager always used to ask her teams what consumers would think about the product if they knew about it. This was a company that produced materials for industry and was many steps removed from dealing with consumers. The manager was very sensitive to the fact that any misstep by a business they supplied could result in liability for her business, too. So, she used to ask her teams about this possible exposure. This technique changed the perspective of the teams. Although it is impossible to determine whether this line of questioning had a positive impact on the business, it certainly sensitized the organization to consider the external perspective at all levels in the business.

Divergent Questions

- Can be used to move to a new line of related thought, but so different that it affects the assumptions used in the discussion

- Allow you to come at the same subject from a new perspective

- Can be used to delay the outcome of a discussion to allow more time

- Inhibit hasty or premature decision making

- Encourage the development of new ideas, new strategies, new processes, and new products

Q: How do sunspots affect the operation of cell phones during the peak of the 11-year sunspot cycle?

Q: How would Sony market fruitcake?

Q: What if we are surprised to find a previously published paper on this invention in some obscure journal?

Q: What else can we make out of okra?

There are many ways to move off the beaten path with new lines of questioning. What the manager must be aware of ahead of time, before diverging, is the following:

- Do you have the time to diverge?
- Do you have a strategy for converging again once the diverging process takes hold?
- Are you prepared to actually go off in a new direction? Avoid using diverging questions if there is no real interest in the end results of the discussion.
- What is the purpose of diverging? Sometimes it is simply to let people vent or explore. If this is the case, the manager should say so to avoid building expectations beyond the intent of the manager to deliver.

56. Convergent Questions

Convergent questions are used to move toward uniformity, to develop consensus, or to move to a decision. Converging brings all the discussion, ideas, and factors together to close out the issue under review.

These questions can end a discussion that has taken place over a considerable length of time, such as for months, or a discussion that has lasted only a few minutes.

Convergent Questions

- Should be used when a decision is necessary or when consensus is required
- Can close a divergent conversation
- Can call for action
- May stop people from equivocating

These are some simplified examples of convergent questions. The situations that call for the use of convergent questions are, in many cases, more complex. The questions might need to be asked in a series rather than as these questions suggest.

Q: It is time to decide on the advertising—do we want to use the beer-drinking gerbil or the talking sunflower?

Q: Fruitcake distribution is different from electronics, so how are we going to solve our problem using some of the lessons we have learned in this discussion?

Q: Now that we have had a year to develop all the test data, is it now time to invest $20 million in this new process?

It's often necessary to draw a discussion to a close but to avoid doing it in a way that chokes off all debate. The gerbil versus the sunflower example actually closes debate. So, the approach taken to converge might be started by the manager by following two different paths of questioning:

1. The manager could ask about the testing criteria as a way off introducing the need to converge, which is often a good strategy to follow. Signaling that you are ending the discussion is often wise. It allows for final points to be made and provides the opposite sides, if there are any, to sum up their arguments.

2. After asking the signal question, a follow-up could be used to make a comparison of two sides (gerbil versus sunflower). If the discussion has no sides, asking for a participant in a discussion to identify the key issues is another way to move forward toward convergence.

57. Redirecting Questions

A redirect question refocuses attention on the questioner's issue of choice and away from whatever the respondent is discussing. It is an effective tool to steer conversations around indirect answers and to avoid confrontations.

Redirect Questions

- May end a contentious discussion or steer away from an inflammatory remark

- Redirect to broader issues when the conversation has gotten too detailed (in the weeds)

- Reexamine previously mentioned details, strategies, or other questions

- Revisit questions (if, for example, new information has been revealed that changes or could change the opinions or decisions already made in the discussion)

- Reestablish a line of questioning

Q: I think I understand what you have said by referring to the prolific wildlife in our area, but I still would like to know how the chipmunks obtained entrance to our clean rooms.

Q: I am interested in how to apply your theory to knee implants as opposed to the robots you are discussing.

Q: Can we return to the question of the cost of this process? Although the benefits are clear, how much more will it cost us over our current method?

Redirecting is also a way to disagree with what was said without actually having to disagree. It also reduces the creep of the unrelated subjects into the conversation. In addition, a manager can use a redirect without exposing his or her personal bias for the information just covered. You might want to dismiss it, or return to it later. The redirect opens these options.

58. Negative Questions

Negative questions have a limited purpose; they are positioning questions. They work similarly to a personal-positioning question except they are asked in a way that negatively positions an issue, a set of facts, an action, a plan, or a proposal.

Negative Questions

- May be used to negatively position an issue
- Can be a way to cut off debate by relegating the issue (action, plan, and so on) to a lower-priority status
- May ascertain negative reasons, assumptions, or logic behind actions, recommendations, or even inaction

Q: Why would you want to enter a market that is so far in decline it is almost dead? (a negative pretense)

Q: Why can't we do this?

Q: Wouldn't it be more cost-effective to leave the air-conditioning system on all day long instead of turning it off at lunch?

Q: You cannot possibly be suggesting that we consider acquiring PDQ?

Q: Wasn't this asked earlier? Is there a reason we are unable to find the answer to this question?

The negative question may also be used in a similar manner to the rhetorical question with the answer already in it. The answer being requested in this type of a question is a negative one that is in agreement with the premise of the question. The management purpose of this use of the negative question might be to chastise, point out the deficiency of a system, or to criticize in an indirect manner.

Q: Have we not done the mailing on time again? (or, even better: Is the mailing late again?)

59. Either/Or Questions

Either/or questions can be used in variety of ways. They can help assess opinions, make decisions, poll a group, force a commitment, or narrow the number of choices.

In the following example, John, the vice president of a business, is speaking to his boss, the chief executive officer. Because of an environmental problem with his business's highest-volume and highest-earning product, John needs to find a replacement in a hurry. His research budget has been increased by 50 percent. At his last meeting with his boss, he had narrowed his choices to four. At this meeting, the CEO expects to hear which of the four John is going to recommend.

> **CEO:** John, why do I see a list of six possible replacement products?
>
> **John:** That's the short list. R&D has indicated that there may be as many as 11 possible replacement candidates.
>
> **CEO:** John, we had agreed to make a decision today on which one was going to be our leading candidate and one backup, did we not?
>
> **John:** Yes, but....
>
> **CEO:** Yes, but I see a list of six.
>
> **John:** One of these other candidates might be the best alternative for us in the long run.
>
> **CEO:** John, if you had to choose either the original list of four on the left or the two possibilities on the right, which list would you work on?
>
> **John:** I would work on the original list.
>
> **CEO:** For our two candidate products, either you are prepared to make a recommendation or someone else will make the choice. Which is it?

John made his recommendation. In defense of John, he was trying to do what he thought to be in the best interest of the business in the long term—choose the product with the best possible outcome. In defense of the CEO, he was doing what was in the best interest of the business in the short term. After all, without a short term, there is no long term.

60. Loaded Questions

Q: So, Mr. Enron, if the company was going to do as well as you projected, why did you cash in all of your stock?

All questions are loaded to a certain extent. If managers follow the guidelines recommended in this book, some thought is always given to the answer before any question is asked. This anticipation loads the question to a certain extent. A *load* is a hidden meaning, an implication of something other than what the respondent may intend, or a method of obtaining a response that might be unintended by the respondent. However, the true nature of the loaded question is to find a way to break down an argument, position, plan, statement, or story.

Loaded Questions

- Are used to get a commitment that the respondent might otherwise not provide
- May be used to force a specific direction in the conversation
- Should be used sparingly

Loaded questions differ from trick questions. They are clearly marked as "loaded." The best example of a loaded question is the one the press asks every president who is ever confronted with a potential conflict. It goes something like this:

Q: So, Mr. President, does this mean you are ruling out the use of force?

Any answer is going to get the president into hot water with someone. Yes, no, maybe, perhaps. These kinds of questions in the political arena are designed specifically to develop a headline. There is little value to the actual answer itself, whereas a trick question is designed to deliver a valuable answer.

61. Trick Questions

Employing trick questions or attempting to expose something by trapping respondents during a business discussion is not recommended. Unless a manager is in a particularly contentious discussion or in heated negotiations where there is little risk in damaging anyone's sensitivities, it is unwise to attempt trick questions. If the veracity of the speaker is in question, probing around factual issues is a better path to follow. Trick questions are entrapments.

Examples of a few tricks and traps are included here to acknowledge their existence, but their use is not recommended.

Trick Questions

- Are hard to use without causing distrust among the respondents (and so it's difficult to find the appropriate circumstances to use them)

- Can be used to test the truthfulness of a respondent or assess guilt

- May be used to discover pretense or uncover resources or references that lack credibility

- Gain commitment from an unwilling respondent

Q: When did you stop beating your spouse?[27] (Trap a person into admitting guilt.)

Q: Now that you are no longer in therapy, tell me what possessed you to select "Water Bag" for the new adult incontinence product? (Admittedly, there is a rather personal preamble in this question, but it does compromise a personal detail that the respondent must either deny or find a way of answering without exposing him or herself.)

Q: Mr. Scopes, you are, of course, familiar with Mr. Darwin's theory that explains why mice do not grow to be as large as refrigerators? (a trick question designed to expose pretense)

The "wife beating" question is commonly found in cross-examination text-books and in jokes about lawyers' behavior in the courtroom. It typifies the kinds of questions used by attorneys when they know they will not be working in a business setting with the person whom they are interrogating. The business environment is not the best place to use this approach.

62. Dual-Answer Closed Questions

Some questions have two possible responses—both are acceptable. Typical of these questions are yes/no, agree/disagree, and male/female questions. Market research surveys use these types of questions to develop information for statistical or qualifying purposes. The answer is important to develop demographic information or to qualify the respondent for further questions in an area of expertise. Any answer within the parameter of the options the questions permits is acceptable to the person asking. These questions are almost always designed to lead to more questions for clarification.

Dual-Answer Closed Questions

- Categorize a person or qualify the person as having expert knowledge

- Establish an opinion

- Obtain an exact answer where the specific details of the answer are unimportant but obtaining an answer is the objective

- Establish a specific position or recommendation

- Reduce equivocating on the part of the respondent

- Set up follow-up questions, such as a *why* question

Q: Do you believe we should consider moving our production facility for Podunk to Pango-Pango? Why?

> **Q:** Which type of response do you think is better, fax or e-mail? Why?
>
> **Q:** Do you agree or disagree with a change in strategy to Plan B? Why?
>
> **Q:** What do you call, heads or tails? (This implies that after winning or losing the call, the respondent then has another question to answer, such as this: If you win the coin toss, do you want to kick or receive the ball?)

63. General Reference Questions to Keep Handy

One basic rule should be followed in all business settings. And, although it seems like common sense, it's worth mentioning before we discuss good questions: *Before asking any question, the manager must know what to do with the answer.*

Thinking of questions is both a conscious and an unconscious act. Developing good questions takes work. However, some shortcuts exist. Keeping a list of questions around to help you in many different kinds of business settings might prove useful. So, here is a list to keep around in case you need them:

What do you mean by that?

What does that mean?

What difference does it make?

What are our options?

How would you decide?

Why?

Can you explain?

Do you know whether there is a problem?

What should we expect?

Have we ever seen this before? Where? When?

How much more is possible?

What are the limitations?

How do we know?

Why do we see changes?

How did it get that way?

To what extent is this supported by experience?

What do you mean by your assertion?

How long have we done it this way?

How much more does it cost?

What is the incremental difference?

What are the alternatives?

What if we tried to approach this from another angle?

What is likely to happen if we are successful in accomplishing this objective?

What should we expect?

Who is taking responsibility?

What if this doesn't work?

What if it works better than expected?

Why are we doing this?

Add to this list if you like and carry it around tucked away for that moment when you are asked to attend a meeting, it's late, you are not prepared, and you need to play an active role in the discussion.

Vary the types of questions that you use whenever possible, as long as you apply them to the appropriate situation. If you want answers—use direct and closed questions. If discussion is needed, then use open or indirect questions. The type of question used will dictate the type of inquiry you are conducting and affect the quality of the answers you receive.

Use of Skills

64. Do You Have a Plan?

Knowing and using the appropriate questions, at the appropriate times, asking the appropriate people, and obtaining what you need in addition to communicating what you want is an extremely difficult job. The objective is to succeed as often as possible, but not to expect success with every interaction. This is a modest objective that takes practice to achieve.

For some managers, the belief that all interactions have been successful is one of the consequences that I have observed in accomplished leaders. In the earlier example where a senior vice president asked a manager to lie to her, her accomplishments over many years had led her to believe that all her interactions and her questions resulted in successful outcomes. For these people, when a business has been particularly successful, habits set in and all of their interactions are viewed as productive. In my opinion, this behavior seems to be a natural consequence of years of good management. However, this often comes to an abrupt halt.

Unanticipated changes in market conditions, unexpected challenges from unlikely competitors, a new business venture, or a change in management in a critical part of the organization may yield problems. The business in our earlier example suffered severely after the interaction where the vice president asked inappropriate questions. In these cases, boards need to step up to the challenge and begin asking questions—different questions from those of senior management.

Accomplishing the modest objective of improving each interaction takes practice. A strategy or plan is the suggested route to take for most formal settings when you will be in a situation where you will be a questioner, interrogator, examiner, detective, interviewer, or inquisitor. This chapter provides insight and guidance into many of the more common management situations.

Managers need to enter business discussions with more than their "wits about them" (if only all managers did have their wits about them). They need a skill set and then some templates on when to recognize a situation that calls for probing, for example, or how to go about testing a new business idea.

A few basic ground rules might help when questions are in order.

General Questioning Strategy for Most Business Settings

1. **Identify the specific kind of situation you are in.**

   Is the setting formal? Are people presenting to you as the primary audience or to others, or is this more of an informal discussion? It is a small step but an important one to consider in all settings where you are functioning as a manager. It is equally important to understand whether you, as a manager in this particular setting, are expected to be there in an "inquisitor" role. If you are just visiting, a polite question to demonstrate that you are paying attention is advisable (however, in-depth probes are not).

2. **Watch carefully for the need to follow up and probe. This is the most neglected area of questioning by managers at all levels.**

   People say things in meetings that often go unchallenged when there are clear and obvious follow-up questions to be asked. In some cases, the time or the setting is inappropriate.

   In one of the marketing jobs I had, we used to prepare and present quarterly reports to senior management. These reviews were designed to familiarize management with the staff rather than to review any aspect of the presentation. Questions were asked to demonstrate interest and to test the ability of the presenters to think on their feet under pressure. These were not the types of reviews where follow-up questioning was appropriate.

   That said, however, I have seen people slip into their discussions a couple of key points that they think need to be made, when they can "get away with it." If you see that happening, you can follow up the next day or at some later, more appropriate time. Just because it was mentioned to senior management during a business review does not mean that it was approved, condoned, or even heard.

118

In some cases, managers will be presented with information that makes no sense, like "it only appears that we had a revenue decline, but this was due to currency imbalance" or "although the scrap rate is high, we believe we will have it cut by 50 percent by the end of the next quarter." These kinds of statements need to be followed up.

In the currency-imbalance case, what the speaker really meant was that there was no imbalance. The forecast was way off the mark, and the young manager was taking advantage of an unfavorable change in rate to explain poor contingency planning in the business. The attempt to create some confusion around business performance was cleared up, after the fact, to the disappointment of the manager.

The scrap-rate comment on the surface looks good, but when questions were asked that peeled away the layers of data that backed up this comment made during a management review, it was discovered that the plant was currently scrapping a full 50 percent of finished product, not scrap produced along the production line. In many plants, a considerable amount of scrap is a relatively normal occurrence as a byproduct of the process, but not in this case. A 50 percent reduction in this particular scrap rate still left this business with an unacceptable level of 25 percent scrap at the finished product level.

3. **Determine a path to take for your questions. For example, follow up for clarification; if unsatisfied, challenge the respondent and probe for details. If necessary, redirect questions back to the main discussion.**

This is an intuitive skill for many managers, whereas many others have absolutely no skills in this area at all. For those people who have had good teachers, or who have an intuitive sense of mapping out questioning strategies, this appears obvious. For others, it takes practice.

Managers occasionally get themselves in a situation where they are led down a path the respondent chooses rather than the path dictated by answers to the questions being asked. I have seen them become exasperated by an inability to get the answers they want, or they take the bait, so to speak, and follow a line of questioning that leads away from the original focus. In either case, a successful obfuscation has occurred—not good for business.

A marketing program piloted by a new manager in one region had resulted in a 10 percent growth in the business in just one quarter. This was a huge increase, almost one whole market share point in that particular market. This explanation was accepted by the VP. It never occurred to him to probe into the reasons for this success. All of it came in one contract. Everyone wanted to be a part of the success, so little probing was done beyond the numbers, and no one wanted to stand in the way of what appeared to be a success. When the program was rolled out to more regions, little growth occurred. Success and failure deserve the same level of attention.

To avoid making inappropriate assumptions, consider mapping out some simple redirecting strategies. It can look like a decision-tree map to make it simple.

Q: That is great news. We moved up a share point, but how many new customers ordered, or was there an increase in ordering from existing customers?

You could have had this question in your mind beforehand, or it merely could have come to you during a product-review discussion. In most cases, data is fairly well known before people get into a room or before the person appears for a meeting or even a phone call. In either case, a brief list of what you will do suffices for detailed directions on how to do it.

A: The increase is due to new business.

Q: Good. So, how many new customers?

This is an acceptance of the response showing respect to the person answering while demanding the same in return. The marketing manager, if he were the person answering, would be putting the best face possible on the data. The probing, which was not done in this example, would have saved a lot of time and money for this particular business.

4. **If your question provokes anger, do not argue. Instead, explain, ask, and redirect.**

We have already discussed using a raised voice to communicate anger. Respondent anger or argumentation should be avoided in most circumstances. In many cases, it is an effective way to deflect critical questions. This is common to HR discussions with employees when a sensitive issue is raised, such as attendance.

Q: WHAT DO YOU MEAN I AM NEVER AT MY DESK WHEN YOU CALL? DO YOU KNOW HOW MANY CALLS I GET EVERY DAY? DO YOU HAVE ANY IDEA HOW MUCH CRAP I HAVE TO PUT UP WITH ON THE PHONE?

The person putting up with crap here is the manager, if she lets this conversation run too long. She must maintain focus on the issue. In this case, the employee was being paid to answer the phone in an environment that provided a lot of personal flexibility. The fact that she was using anger to deflect her supervisor's questions clearly demonstrates the problem. This kind of response is found at all levels of business.

A senior vice president/divisional general manager in a large multi-national company, when asked by the CEO to explain a hard-to-believe financial turnaround by one of his business units also responded with anger.

VP: YOU OUGHT TO HAVE PEOPLE IN THESE JOBS YOU TRUST. THE BUSINESS IS DOING WELL. THE MANAGEMENT TEAM HAS DONE A HECKUVA JOB, AND I THINK YOU SHOULD BE CONGRATULATING THEM!

CEO: Thank you, Bert, for a proud defense of your business and your team.

This was a public meeting, and old Bert thought he had gotten away with a masterful performance of deflection, which he had displayed on many occasions. The CEO let others ask the appropriate follow-up questions. Bert later chose an early retirement when an audit, ordered by the CEO, revealed irregularities in a number of financial areas.

5. **Avoid pursuing unrelated and unimportant details.**

An agriculture business had developed a new fishing bait using waste product from one of the manufacturing processes. It had a number of positive environmental attributes. It removed a solid waste problem and turned it into a nonpolluting product that could generate sales from a relatively low investment. It was nontoxic and biodegradable.

After sitting through a two-hour discussion of the technical problems of producing the fishing bait, the business manager simply turned around and said, "So, tell me, do the fish like it?"

If the fish didn't like the bait, solving all the manufacturing problems would not help the product one bit. As it turned out, the fish were not biting.

6. **Close an inquiry with converging questions.**

This strategy allows the lose ends to either be tied together or recog-nized as such, setting up future discussions. Unless the discussion is going to be continued, it is suggested that some convergence be the objective of discussions that produce much difference of opinion.

Convergence does not mean agreement. It just means that the argument is being brought back to a central point.

Q: How many choices do we have for odd lot materials to list for sale?

Q: Because it appears unlikely we can afford to support the sale of all of those products simultaneously, is there some criteria we can agree on that will establish a priority for accepting and then listing product for sale?

7. **Keep an open mind.**

An open mind is all that can be asked of anyone in any management position in business today.

Q: What do you mean by "I don't get it"?

A: You don't. Do you have any idea how we do our jobs in the division? Do you even understand what it takes to compete for promotion and bonuses in this organization? Did you bother to ask any of us before you made your promises to senior management?

And so, the business manager in this case sat through a full three hours of complaining by his staff. She would stop them every once in a while to ask a clarifying question, but other than that, she sat and listened attentively to what was being said.

They were very unhappy with a hands-on manager after having lived through a series of managers who were absentee-leaders. The last two people spent more time on the golf course than they did with the staff. This manager needed to wake up her moribund team and did it by putting forward an aggressive forecast to her management. After the three-hour rant, the team realized that they could do nothing to change any forecast already made. The manager also realized that this team had little motivation for achieving anything other than what they had been doing during the past few years. No one on this staff had received a promotion in level or a substantial bonus for good performance. They were angry, but not with her.

So, after changing the incentives, the business team ended up achieving higher-than-forecast results. They were amply rewarded at year's end. Had the manager decided to push the plan forward, she never would have learned about the reasons for mediocre performance and would have wrongly concluded that the previous managers had just not put enough effort into the job.

8. **Do not ask too many questions.**

   One of my good friends had been a corporate planning director while he was being groomed for a more senior position. Unfortunately, he developed a habit of asking too many questions in every review he attended. He felt it was his duty to expose as many soft spots in the business as possible to improve strategic planning. This did not work well for him. Although he became widely recognized for his insight, he was dropped from further promotional consideration because he simply had not learned when to stop.

   Earlier in this text, we discussed the downside of asking too many questions. A person might be able to get away with this once or twice, but not as a regular business practice. If continuous questioning is required, it should be moved off line to a private setting. Then, it is important to maintain focus on asking questions until you get the answers needed and nothing more.

9. **Listen for the complete answer. Do not interrupt (unless for some egregious error or deception).**

   This is a common courtesy not always followed by managers, particularly arrogant ones. They appear to behave as if interrupting people is an entitlement. Some of them will continuously badger their employees with questions before they have even had a chance to consider answers to the previous questions. This is both bad manners and bad management.

10. **Stop when you are finished.**

    The last rule might seem obvious, but how often have you seen a manager continue to drill into a topic when it is no longer necessary? If you are uncertain about knowing whether you have finished, use a pause or a silent question. People often respond with additional information in these situations if they think it will add to the discussion. If further information isn't forthcoming, you might be at the end of that particular line of questioning. You can always return later if another issue arises that needs attention.

Managers also often forget to use a question to redirect respondents back to the major topic of discussion. They tend to "order" a return if they remember. A question is much more effective because it requires the conversation to be focused on an answer of interest to the manager rather than a general topic.

The general direction of a discussion may look like this:

Example: Business Plan for a New Market

Q: (**Clarification**) Do you have a reference for those market-growth projections?

A: The numbers are from an article I read on the web.

Q: (**Follow-up**) What is the specific reference?

A: An article from the *Podunk Journal* dated April 1.

Q: (**Start to probe**) This isn't good enough. The information is suspect. (Use a negative question to get the point across about how you feel about the reference without the need to have a major discussion.) Isn't that the "journal for little-known and less-cared-about knowledge?"

A: Yes.

Q: (**Still probing**) Who was the author?

A: Dr. Adam Smith.

Q: (**Still probing**) Did you try to corroborate this with any other source?

A: No.

Q: (**Converge**) Can you find another source after this meeting is over?

A: Yes.

Q: (**Redirect**) If we accept this projection for the moment, how many market segments shown on your last slide does this represent?

The reason the manager is at this meeting is clear—this is a discussion meeting to vet a new market development plan. It is not a decision-making setting. If it were and the data were as suspect as it is in the example, the manager might move to directly challenge, or even provoke, the respondent. Another option, for a cooler-headed manager, instead of provoking is to use hypothetical questions.

Q: I am very concerned that Dr. Smith's work might be a little out of date. What if he were wrong by a factor of two? What affect would that have on your projections?

A challenge would look a little different and accomplish another purpose—a performance review.

Q: Adam Smith has been dead now for a couple of centuries. I think his data is a little out of date. What do you think?

This is generally not a good idea unless a person is being particularly recalcitrant.

Specific strategies are made up on the spot. The generalized model is not. These kinds of models are built by stringing together the types of questions mentioned earlier in the book with the questioning platforms described in this chapter.

The premise is that managers should begin by assuming their own ignorance when asking questions rather than banking on what they know. That is the foundation of a solid strategy for asking questions; it is built on what is not known.

Find a Strategy

A *strategy* is a plan to reach a specific objective. Most questions that come to mind during normal business discourse are part of the more generalized plan of the business—they are part of the process that helps accomplish the tasks to contribute to the goals of the organization. However, it is vital to have a plan when your questions are part of a probe or investigation or for some other more focused purpose.

Explicit strategies are recommended for these situations:

- Conducting a probe
- Following-up questioning
- Evaluating ideas, plans, products, problems
- Investigating a specific situation
- Controlling a discussion
- Asking hard questions
- Finding fatal flaws

List what you expect to learn and a few specific questions that will lead to the required information.

65. Follow-Ups and Probes

Follow-up questioning is a normal part of conversation, and probing strategies should be considered an extension of that approach.

For mistakes in reasoning, analysis, or by omission, good follow-up questions can bring out the cogent details in the normal course of questioning. It is always a good idea to follow up when observing mistakes instead of merely pointing them out. The objective of using follow-up questioning is not to set the record straight; it is to focus on information important to decision making, financial considerations, accounting, or on how business is conducted.

There is so much in business that is based on conjecture that information about markets is usually a view of a constantly moving target. Businesses need to focus forward to move product, improve sales, fix problems, adopt or adapt new technology, and find new opportunities. The process of ensuring accuracy or making certain that the record setting is correct is usually focused on the financial and accounting aspects of the business. Probing questions when the numbers do not add up or when things do not make sense are always important to businesses.

Many extraneous details are unnecessary to follow up on, even if wrong, provided they have no bearing on decision making or on costs or how you do business. However, the glossing over of vital information or even sidestepping important issues is the kind of thing where follow up is necessary.

Manager: Tell me, Mr. Enron, just how did you grow from a start-up business with no revenue in June to $4 billion by the end of the year?

Mr. Enron: We held bake sales until we were able to get enough windmills operating in the North Sea to sell electricity to South America.

Manager: Speaking of baked goods makes me hungry. What are we having for lunch?

Asking about lunch is not a follow-up question, yet this kind of lack of follow-up occurs often. This particular situation should also move quickly from a follow-up phase to a probe. Both kinds of inquiry strategies are discussed in the next two sections.

66. Follow-Up Questions

Follow-up questions continue the discussion in more detail. They can also be used to follow up on events or conversations of previous days, or to raise questions that are more current.

Previous discussions do matter. It is important to keep in mind that conversations are like e-mail messages, except they don't carry any string with them as they wend their way around people. Follow-up questions can be used in this context, as a way of continuing a discussion, or, more commonly for managers, as a way of gaining a better appreciation of subjects under current discussion.

Listed in the following table are general-purpose follow-up questions flexible enough for most occasions. These questions can either be open or closed depending on what the issue is.

For example, if the name of a reference is requested, a short closed question is all that is usually needed. However, if a lot of uncertainty is expressed about a recommended course of action, a more open "What did you mean by that?" inquiry is helpful. This style of questioning allows the story to continue in more detail without introducing bias or passing judgment.

Indirect questions are not very useful as follow-up questions. Negative questions are also less useful than other types of questions. Negative questions have been kept out of this discussion because they raise more issues than they resolve. Management should normally be looking for closure. That is the peculiar and necessary role of business. Negative questions do not necessarily bring the kind of closure usually being sought in normal situations. Although necessary in many contexts, such as in a cross-examination, their use in business tends to politicize many discussions unnecessarily. It is much better for a manager to speak directly to an issue expressing his or her opinions, beliefs, and thoughts. Doing so maintains momentum.

Type of Question	Potential Use as a Follow-Up
Direct	Direct answers are needed.
Open	Telling the whole story may be important.
Closed	The typical type of follow-up question.
Pause	A good way to fill in a missing detail.
Silent	A way to encourage the speaker to continue.
One-Word	Quick follow-up such as "When?" "Where?" or "How?"
Hypothetical	Allows for the inclusion of other information.
Clarifying	Another common follow-up question.
Either/Or	Use when two options or conditions are present.

General Purpose Follow-Up Questions[28]

This is a quick reference list of the kinds of follow-up questions that can be asked in a variety of situations. They represent avenues of additional inquiry that may be open, or offer a manager, board member, or others a mechanism for engaging in discussion without appearing as if there is any ulterior motive. Most managers I have observed generally have habit follow-up questions in addition to whatever habit questions they might acquire as part of their practice of management. By using a different tool—a different question—a new perspective can be gained. It is also equally important for managers, in particular, to demonstrate a number of alternative approaches to asking questions.

What do you mean by that?

What was the result?

In what way?

How did that come about?

Were the conditions different?

How were the conditions different?

When do you expect this to happen?

How often does this occur?

Which way are you leaning?

Can you cite some examples?

How did the glitch you mentioned yesterday affect the operation today?

Then what happened?

What do you mean by that?

Who else?

How much?

When?

Where?

How did that happen?

Have you any alternative theory that will meet the facts?[29]

What Specific Situations Call for Follow-Up Questions?

A lot of us sit through meetings where, for one reason or another, good follow-up questions are not forthcoming, even when we believe they should be. Consider asking follow-up questions when confronted by any of the situation descriptions outlined here:

- **When the assumptions are not clear**

  Q: What exactly were your assumptions for projecting that this product will capture 20 percent of the market?

  Q: Although it appears to be a good idea, what was the basis of this decision?

  Q: What do your assumptions have to do with chipmunks?

- **When the answer to a previous question is unclear**

  Q: When you stated that the exchange rate was lower, lower than what?

  Q: I am uncertain exactly what you meant. Could you repeat the answer for me?

  Q: What did you say again?

  Q: Can you explain in more detail?

- **When a different question is answered than the one that was asked**[30]

  This issue is critical. People often respond to questions with an answer to a different question. Among the reasons people do this are they are attempting to change the subject, redirect the discussion, or avoid an embarrassing answer; or they are unprepared for the issue you have raised and are attempting to deflect it.

  Managers who have undergone media training are instructed in how to do this. Executives facing the press are often asked questions publicly that they cannot or would not ever answer. The executive might get away with this in front of a reporter or TV camera, due to the limited amount of on-air time. (Because in these situations, the important issue is to ask the questions.) There is a certain amount of expectation that the answers might not be direct because the agenda of the person being interviewed may be to make a completely different point than the one the reporter wants to make. Although the person might get away with this strategy on TV, no manager should ever fall prey to this distraction when asking questions.

This may also be a signal to start probing. The first step in response to the misdirected answer is to pose a follow-up. If the question is avoided again, start probing. There is always a reason for a misdirected answer. These are not casual errors in discourse.

It isn't always necessary to restate the question in an interrogative format to return the conversation to the direction you want. A direct reference to the question just asked either by saying you want to readdress the question or by referencing the subject matter is sufficient.

Q: Okay, I understand what you are saying, but I was asking a different question.

Q: That may be, but I still want to know, why should we buy Greenland?

Q: That's very interesting. However, the question is why did you spend a billion dollars on balloons?

Q: Aside from cockroaches, our interest is in finding the answer to the termite problem.

Note that not all questions sound like interrogatives. The first and last statements above are really questions hiding in comments. They are responses to statements made by others that were evidently not answers to the questions asked.

- **To ask for references or for sources of information**

  There are always varying levels of expertise in any meeting, particularly those on technical subjects. I have heard managers leaving a room mention that some aspects of the subject were unfamiliar to them, and it is in these situations that this question is useful to ask in the open setting.

  These kinds of questions seem to occur most often in meetings discussing medical topics. Audiences or meeting participants are very open about their interests in learning more about a particular subject when health is in the balance. Managers, on the other hand, are cast as generalists in many of their assignments, expecting the knowledge that carries the business to be possessed by the staff. The veracity of the data is taken for granted. Also, a general belief pervades most businesses that opinions or assumptions based on inference are normally stated as such. No one likes to be wrong in business because the consequences can be career ending.

So, asking for a reference or a source does not mean that the subject matter is of such extreme interest that there will be follow through—it might mean that the information provided may be worth challenging. Data is often quoted, trends are cited, and government actions referenced in many discussions of importance. Asking for references is not critical in all circumstances, but on occasion, when the facts are surprising or unanticipated given what may have previously been thought, it is always good to ask for the source. In addition, it's also wise to consider asking for those references that support the prevailing wisdom, too.

Q: Could you provide me with additional references?

Q: I would like to learn more; where can I find a couple of good articles?

- **To ask for contacts**

  Q: Who are your contacts in Kandahar?

  Q: Which of the people you mentioned provided this information?

  Q: Can you name names?

  Q: Who are they?

  Q: I want to hear more on the subject of musical cereal; who should we contact?

  Q: Are there others who believe as you do?

- **If there is a need to reference a previous discussion**

  Q: Can you explain this in the same way you did yesterday?

  Q: Since you mentioned camels before as a primary means of locomotion, just where are you planning to do this market research study you are proposing?

  Q: How is this different from our previous discussion?

- **When you need familiarity with an issue**

  Do you have enough familiarity with the subject? Ask yourself the "What if I were asked about this in court?" question to determine whether you know enough.

  Q: What else do we need to know before we make a decision?

  Q: What questions do we not yet have answers for?

  Legal When a reference may be made to a law as a reason for doing or not doing something, managers should consider asking about the law. What is it exactly? How does it affect you and your business in this situation?

Q: What is that law (regulation)?

Q: What are the implications?

Q: Is it possible that we need additional legal resources?

Q: How certain are you?

Regulatory The same issue applies here as with a law.

Q: What are the regulations, the agency, and the ramification of following or not following the regulation?

Q: Are there any other regulatory agencies that have jurisdiction here?

Q: How do we manage our compliance?

Q: How is it that our Pango Pango manufacturing plant is regulated by the city council of Podunck?

Economic and financial If you are unaware of what "generally accepted accounting practices" are, the next time the phrase is used, ask. As we have seen in the popular press in articles about many accounting practices of large corporations, this phrase has been used to cover practices that may be generally acceptable—acceptable enough to land managers in jail.

Q: What do you mean by "generally accepted accounting practices"?

Q: What does this apply to?

Q: How will this affect us?

Q: Who is the guy Adam Smith you keep referring to?

- **Data is inconclusive**

  Q: How soon will we have a better idea about the data?

  Q: Are there any other tests that should be done?

  Q: Should additional market research be considered?

  Q: How can we find the data we need?

An equipment manufacturer was about six to nine months late in getting a product to market—well behind its competitors. Business management wanted to offer the equipment for sale early, before the product was actually ready. So, they asked their attorney if they could take contract orders for delivery at a later date. "No" was the answer.

Follow-up questions were not asked, and the attorney was not invited to the business team meeting to explain her answer.

Two months later, as the product team was making plans for attending a trade show—a show where the competition was introducing new products and they would have none—the marketing group decided to hold a meeting and invited the attorney. She explained that, although contracts should be avoided, in her opinion, customers could reserve product customized for their specific application in advance.

At the trade show, with only a model mock-up of the product, the company was able to interest enough customers that they ended up leading the market in new product orders when the product was released.

Follow-up questions should be asked whenever all information necessary for effective decision making is lacking clarity or when all participants have not had a chance to express themselves.

67. Probing Strategies

Probing may be defined as aggressive follow-up questioning. However, you are not necessarily just interested in keeping a continuous line of discussion going. Probes are used to look for something other than what the discussion, the paper, or the message has provided. You probe when you encounter potential deceit, defensive behavior, half-truths, challenges, misdirected answers, dead experts, and any number of other conditions likely to occur.

Remember the manager who asked questions with the answers in them, and then argued when his staff tried to tell him he was wrong? I saw him in action off and on for about two years. Not once did I ever see him probe any topic. Even if he disagreed with the information presented to him, he dismissed it as "irrelevant" or just plain wrong. Probe to avoid becoming this myopic.

By the way, his business unit failed and, unfortunately, he was actually promoted to a new position of responsibility where he could ruin another business. He didn't disappoint.

A complex mix of objectives surrounds probes. The reason a business manager is doing a probe in the first place, rather than asking simple follow-up questions, is to move the discussion from a straightforward inquiry to an investigation.

Questions Best Suited for Probes

Type of Question	Potential Use
Direct	The probing purpose of questions must be unmistakable.
Closed	Avoid misdirection by keeping questions closed.
Filter	Remove information of no interest to the probe.
Provocative	Issue a challenge.
Leading	Leading may be appropriate in some circumstances.
One-Word	The use of *why* is often most effective.
Double-Direct	It is a leading question that requires a thoughtful answer.
Hypothetical	Use for uncovering missing information.
Redirecting	Use to deal with misdirection from any respondent.
Loaded	For contentious situations.

Launching a General Probe

This list represents launch questions—the road into your pursuit. They are general enough in nature to fit a wide array of circumstances. Probing generally takes longer than following up. For example, when dead experts are used as resources for a business case, you not only need to find living experts, but the manager also needs to understand why the dead experts were cited in the first place.

A continuous series of follow-up questions constitutes an exercise in probing. It can be relentless. The questioning can change directions, and it can be discontinuous if the manager feels that the respondents may be disingenuous in any way.

Who else can we check with?

Do you know where to get additional information? Specifically where? And what does it say?

Why was this particular expert chosen?

Who else uses this...relies on this expert?

Why?

What do you mean by that?

Is there anything else we should be aware of?

What else? Do you have a list of concerns?

Specifically, what should we be concerned with and why?

Is there anything about this that keeps you up at night? (A habit question of an old boss of mine and a good one, too. He had a number of different versions, but he used it to elicit many issues that had not been mentioned in "regular" discussions.)

How can you be certain?

What can we not rule out? Why?

The purposes of probing are varied. Probes are conducted, for example, to determine the credibility of a speaker, the importance of an issue, factual details that have been ignored, or because of a gut instinct about needing more and different kinds of information—searching for something other than what is being presented or discussed with you. Probes are suggested when the situation calls for it. Here are some of those situations when probing is advisable.

When Do You Need to Launch a Probe?

- **Violates the laws of gravity**

  In an earlier example, the growth of a business from zero dollars in revenue to $4 billion in six months "goes against gravity," as an old colleague of mine liked to say. Common sense tells us that this type of growth is so improbable that it approaches impossibility. If this kind of growth were to occur without an acquisition, you had better get out of the way for an investigation.

  Q: Why did you forecast everyone in the world buying one of these?

  Q: It was quite windy that day, but how likely is it that the wind caused the coffee to spill inside the building and all over the server?

  Q: Dr. Deleon, how do you know you discovered the fountain of youth formula? What evidence do you have? Who else has tested this? What were their results? What do you mean they are now too young to answer?

- **Dead or unavailable experts**

  Dead experts are a dead giveaway that probing for better references is needed. If an expert whose knowledge is critical to whatever business case it at hand is unavailable for a period of time that exceeds your decision-making timeframe, probe for another expert.

  Q: Who, other than Adam Smith, can we contact on this theory?

  Q: Why is it the only expert in the world on this is in Antarctica when we need her?

  Q: Yes, jail is a difficult place to hold a meeting, but another year seems a bit longer than we can afford to wait, doesn't it?

  Q: Who did you pay for this information? How much? Did you get bids from other providers? Who and how much?

- **When the respondent continues to ignore a follow-up question and answers a different issue**

  If this happens, move from follow-up to probe. There is always a reason for this strategy by respondents. Your objective when you shift to probing is to avoid being judgmental. Just because the behavior before you may indicate a problem does not necessarily mean that there is one. Maintaining objectivity until you get all the facts you need is important to the questioning process.

  A direct inquiry approach is best. It avoids wasting time. A direct approach also pays off in the future. You are less likely to be faced with misdirection answers from others.

  One additional point is worth noting here. You can change your style from a facilitative (kindly) management approach to a more control-oriented or prosecutorial manner (a more adversarial managerial style). The person knows exactly what it is he or she has done, and you must indicate that you want answers. Obfuscation is for politicians and diplomats, not for businesses.

  Don't take the bait by following this rabbit down the hole. Persons good at this strategy will usually issue an enticement in a misdirected answer; it's likely to be some detail that is well known to be of critical interest to the inquisitor. I have seen managers fall prey to this and realize hours later that they didn't get the answer to the question they had really asked.

I saw a regional sales manager practice this with his VP of sales. The manager had arrived at the home office just in time to attend a meeting where his interest was a lack of proper accounting of about "a hundred thousand dollars."

VP sales: Could you review those numbers again for me, Al?

Regional manager: They bother you, too? You know another thing that is an even bigger concern to me is the Simpson's. They are the largest customer in the country, and we just learned yesterday that they are considering canceling our agreement. Is there any other incentive package we can offer?

How could the VP not return to his original line of questioning? Easy. The Simpson account was huge, well over 20 percent of U.S. sales. Although he might have been able to see through the ruse, he was unable to resist the possibility of the reality of a problem with this account. It could be that the VP of sales was implicitly going along with this misdirection. Asking tough questions of the people you work with every day is often difficult. This is particularly true when you have known them for many years. There is no way to know.

A routine audit of the books disclosed serious accounting irregularities in the preceding case. A change in regional management cleared up the problems and wizened the VP of sales.

The suggestion for responding to misdirection strategies is to maintain focus and keep your probes direct.

Q: Did you understand my question?

Q: Why are you answering a different question than the one I asked?

Q: What makes this information relevant to what I am asking?

Q: Can you repeat what it is I am asking?

Q: How can I be clearer about what it is I am looking for?

Q: What is it about my question that you don't understand?

Q: Yes, that's a concern for me, too, but how much did you say you spent?

Q: We will cover that if time permits, but let's return to my question. What is your answer?

Q: Why are you having a problem answering the question?

- **When the answer is incomplete**

  This, too, could be an attempt at a subtle strategy for avoiding the question. I have seen this happen in meetings with CEOs, for example, when the respondent knows that the time is limited and tries to avoid completely disclosing details of an issue.

  Q: Yes, I'm glad that the mess is cleaned up, but I need to know the whole story. Just how did the chipmunks get into the clean room to begin with?

  Q: Your data does show that all the sinks we sell are of the best quality, but I need the whole question answered—what is the quality of all the sinks we manufacture? We scrap how many? How long do you think we can remain in business with that rate? What's being done? What's the plan? Who's responsible for implementing it?

  These questions reflect an actual situation. A business manager had successfully dodged his VP and the CEO on this issue for years. Literally half of their product line ended up in the scrap heap, but the margins on the finished product were so high that no one paid a great deal of attention. They sold into the luxury end of a market, and no one ever mentioned the scrap rate. This lasted until competitive pressures forced attention to the problem. The business eventually solved the problem and fixed the product line but sacrificed earnings to overcome years of inattention.

  Q: I know everybody here liked the ad campaign, but I wanted to know how it tested with consumers? What are the results from market research studies? You did do market research studies with real customers, didn't you?

  By the way, a manager's voice that goes up at the end is likely to get a more open response.

- **Conflicting information, discrepancies, factual errors**

  A CEO whom I watched in action had a very effective method for dealing with conflicting information. He would state the discrepancy. This is an effective approach because not all discrepant information about a subject is coming from the same person. Information often arrives from a number of different sources, and it makes a better-understood query to state the conflict to the person being questioned.

  Q: Joe, the message from your team yesterday was that the project would be ready on time. However, today you have just indicated that a delay is likely. Can you explain the difference?

Q: How could we go from 3 possible reasons for the problem to 11?

Q: Either the duck got into the copy machine all by itself as you suggest, or as Wilson explained, someone put it in there. Which is it?

- **Red herrings**

  When an answer is irrelevant, it's time to probe. Once again, avoid taking the bait by pursuing the subject, but acknowledge the lack of relevance.

  Q: Yes, branches are falling from the trees due to the drought, but why have our sales have been dropping like those branches?

  Q: How did a herring get into the last batch of paint? It's unimportant what color it was.

- **Equivocates**

  Equivocating is the use of words with multiple meanings. This can allow a person to take a position on either side of an issue. Also included in this category of signals that scream "probe now" are behaviors characterized by beating around the bush, waffling, fudging, and stalling. Straight questions, once again, may provide assistance.

  Q: We all know that potatoes contain healthful substances and that chicken fat does taste good. Nevertheless, can you explain how much of this healthful food aspect is lost when the potatoes are deep fried in the chicken fat?

  Q: Even though the regulation is there for a good reason, and you are correct, we need to document our decision in either case, but we need a specific recommendation on whether we file a new application. What data do you have? What studies need to be done? How quickly can these be done? Who will do them? Is there a reason for your hesitancy?

- **Lacking facts, or lack of evidence to support the claims made in answer to your question**

  Q: Where exactly will this 1,000-store mall be opening?

  Q: Why was our data submission to the FDA rejected?

  Q: What do we not know about this project that we should known?

- **Answers that reflects wishful thinking**

  When the answers to your questions include a lot of "We hope so," "We wish it would be," and "We are encouraged by signs," consider probing. Do this if for no other reason than to avoid the fate that Ben Franklin ascribed to those who engage in this kind of thinking when he said, "He that lives on hope, dies fasting."

  Q: Could you define what you mean by hope?

  Q: What data do you have to support your wish?

  Q: How much faith are you putting in "hope" and how much data is available to support that? Who supplied this data? What are their interests in this project?

- **Constant use of hyperbole**

  Q: When you answered that we will get a billion customers overnight, exactly how many customers do you have in your forecast for the end of the year?

  Q: Although we appreciate the expression that the new drug will change the way the world thinks about medicine, exactly how will this happen?

- **Airtight answers**

  Probing is also recommended when answers to all questions are closed ended, meaning that there are no potential problems, concerns, or issues to worry about.

  Q: Is there anything about the project that keeps you up at night?

  Q: What if something were to go wrong? How could we explain it?

  Q: I understand that there is no scientific way possible for our fertilizer to smell, but what would have to happen for an odor to be present? How much of the state would we have to evacuate?

  Q: Even though you dismiss the possibility, is there a chance that any of us could go to jail? What would have to happen? Give me a list of issues that could precipitate an investigation?

  Q: How solid are your projections? Are you willing to bet your bonus?

- **The "two false options" gambit**

  In some rare instances, particularly when a business is looking for someone to blame for a problem, a choice is set up for the manager. The manager is presented with a choice of selecting between two options, both of which are false.

In one particular case, a product complaint had come from a very influential customer, and the service representative, desiring desperately to avoid any possibility of blame, offered his manager two possible explanations for the problem.

"Either manufacturing hooked up the power cables to the phone jack or the customer plugged the unit into a DC line." There is no way a manager should accept either of these as options without probing around the problem just a little.

Q: What other alternatives are there, and don't tell me that there aren't any?

Q: How many times has either of those occurred?

Q: What conditions would cause us to look at these options?

Q: Pins and bullets both have the capability of puncturing balloons, but what do you really think punctured the balloon at 25,000 feet?

- **Begging the question**

  When the reason in support of the answer is produced within the answer itself—when a conclusion is assumed without proof—it's time to probe a little. Once again, the best approach I have seen is to state what it is the respondent has identified as the answer, and then probe the part of the answer that is unsubstantiated.

  Q: Your answer is correct; grass can and does turn brown in August in that part of the world because of drought. What evidence do you have that it was not our fertilizer that caused the problem?

  Q: Baldness occurs all the time, but not all at once. What is in our product that could have possibly caused 10,000 people to go bald in 2 days?

- **Gut instinct**

  Follow your gut. It might not lead to any particular problem or issue of concern, but it's worth developing any line of questioning whenever a feeling of uncertainty, suspicion, or inquisitiveness comes along.

Probes are important to think about whenever you think about following up on answers you are dissatisfied with.

68. Does the Manager Need to Control the Conversation?

Attorneys use questions to control testimony. Managers can also use this method of control, but under specific circumstances. The conversation-control strategy requires that a series of questions be prepared, preferably in advance if possible. If not, then at least you need to be a question or two ahead of the conversation as it's taking place. Without a series of related questions, many conversations may get out of hand.

A manager, to control her team of unruly engineers, had a long list of questions she would resort to whenever her meetings got out of control—which happened quite regularly. It seemed that the electrical and software engineers were constantly haranguing the mechanical engineers. This conflict resulted in furniture tossing.

This was a new business start-up operation in an old warehouse that had been converted into a kind of service center. The furniture was plastic lawn stuff, and most of it had been damaged in some way long before the arrival of this social experiment. However real the anger appeared to be, no one was ever hit by a chair. Tables were also overturned, but this was the routine way in which the meeting was brought to a close.

Someone would say: Well fine! If that's the way it's gonna be....

The folding conference table, which looked as if it had been left by a catering service too fearful of returning to retrieve it, would then be pushed over. This signaled the end of the intellectual interchange. The discussion inevitably would turn to lunch as the group left the room.

The table-toppling behavior was tolerated by the manager because of the creative nature of this particular combination of people with specialized talent and problem-solving skills. She would sit at a desk at the head of the room and fire off questions as if the mayhem in front of her wasn't happening, but she knew her team. They really were interested in answering the questions, and this is why all the furniture tossing never caused injury or damage.

Control strategies are useful under these varying conditions:

- To establish order with an unruly group
- To deal with continuity problems and the difficulty of staying on subject

- To drive for decision or consensus—when a group or team is not moving
- To meet time constraints
- To obtain all information if the person or group will become unavailable
- To launch probes
- To examine data before moving on

Questions to Use When Employing Control Strategies

Type	Possible Benefit
Direct	Communicates purpose, maintains clarity
Rhetorical	Reminds people to focus
Closed	Moves discussion toward answers
Hypothetical	Determines whether a probe is necessary
Convergent	Goes for uniformity or a decision
Redirecting	Brings back the discussion to the question

Always look for a method whenever you are confronted with madness.

69. Strategies for Asking Tough Questions

A tough question is one that makes the respondent uncomfortable. Some managers find it difficult to ask these questions. This isn't good for the business. What happens in these circumstances? The business suffers. A mistake or even a deception may continue unabated.

When asked, these questions are often taken personally—because they are personal. A question is being asked of a person to knowingly cause that person discomfort. All of us recognize this and take it personally. It's a natural response. Therefore, the first order of business for the manager is to remember this: When you have to ask a hard question, try to remember it's "just business."

I have had to ask tough questions of employees who suffered from addiction, were suspected of stealing, were accused of sexual harassment, and ones who had a host of other serious problems. These are outside the purview of this book. When a legal issue arises, you need the advice of professionals before taking action—if you have a chance. If you are caught by surprise, as I was on one occasion, you might choose to use the strategy outlined for surprises that follows the "tough" question discussion.

Employing a prosecutorial approach to asking might be necessary under a variety of circumstances. A probe may be necessary. Probes take you off the path you are on, whereas these questions help retain the initial focus of the discussion by asking the question needed and then moving on. If you discover a more serious situation is evolving, start your probe.

Good preparation or familiarity with both the respondents and the subject matter is required; otherwise, your strategy could come unraveled.

When to Employ Tough-Question Strategies

Example Situations	Question Strategy
Erroneous data	Discredit data questions.
Faulty conclusions	Question the assumptions.
Questionable references	Probe for credibility.
Gravity law violation	Probe for reasons why.
Improbability	Attack assumptions.
Intransigence	Probe for reasons why.
Suspicion of bias	Ask directly.

There is a risk of damaging relationships when hard-line questions are levied in a discussion. Their use represents a calculated risk for the manager. This strategy requires that the manager has a good working knowledge of the participants.

Q: Jane, why should I believe that your division has $1 million in earnings when the total revenue of the company is only $1 million?

Questions that target fabrication or questionable data do have a downside. They're relatively transparent unless the manager is extremely skillful. They could communicate lack of confidence in people who might not have realized there were problems with their data.

I have seen managers who trust no one, and their questions always indicate this type of an approach. No one takes the inquiry too personally under those circumstances. That's the only good news from the lack of trust strategy. The manager garners no trust either. The group converts from a working organization to an armed camp. There are also managers who should perform veracity tests when they don't. These folks are continually taken advantage of. The business suffers in either of these extremes.

Try any one or more of these approaches before you start to probe for more details. Answers may pop out, making a probe unnecessary:

Who performed this analysis?

Who else has seen this?

Can you provide me with references for that information?

What were the assumptions, and how were they tested?

What did our attorneys have to say about this?

Whom did you speak with?

Would you recommend that I speak with this person because I have an issue with their data?

Would you stake your next pay raise (or bonus) on that? (This tests a lot more than truthfulness.)

What are the assumptions?

Why do you feel this way?

What are your biases in this case?

What is your personal opinion?

Questions Useful in Tough-Question Situations

Type	Use
Direct	Clear purpose.
Closed	Looking for answers, not stories.
Provocative	When challenges might be useful.
Leading	Helps probe an area of interest if necessary.
Silent	Creates a silence that some people try to fill.
One word	*Why* is very effective.
Redirecting	Often necessary.

The surprise that happened to me one afternoon, a few weeks into a new job, was a call from one of our large West Coast accounts. The gentleman on the other end of the phone was so angry he was incoherent. However, I was able to learn that his assistant had been on the receiving end of serious harassment from one of my new staff members. I immediately recalled this person from our California office to the East Coast, which gave me time to get the advice of legal staff and to assemble an appropriate set of questions.

Contracts were not canceled as threatened, nor was court action taken. After a series of questions to the manager, he admitted to the problem and was demoted, reassigned, and required to enter a counseling program. The questions were personal. They elicited ugly responses at first, and it was unpleasant for us both.

70. Mounting Challenges

A number of different types of questions can be used to challenge the person or persons you are addressing. Why do you want to challenge people? Doesn't this lead to a confrontation?

Yes, although not always. A challenge can be brought forward without causing a confrontation as long as the line of questioning is viewed as a challenge to the substance of what is being presented rather than to the person presenting it. *What if* questions work well in this type of situation:

What if our competitor produces our product?

What if we had lost money this last quarter?

What if customers refuse to purchase unless we cut prices in half?

What if we need to reduce the number of employees by half?

What if we lose all telecommunications?

What if we are wrong?

What if our assumptions are based on faulty information?

This is a less-confrontational manner to challenge, instead of asking for proof that the respondent is "right," which I have witnessed on occasion and which turns the session into an inquisition. Asking *what if* also encourages creative thought. It lets the organization know that the manager is willing to reach outside of the confines of the usual business discussion to listen to new ideas.

What if questions can also be used to shut down discussion and control meetings in a way that discourages rather than encourages debate. A general manager was listening to a presentation from his technology development organization. The research director, after listening to a discussion of their new technology versus that of a competitor, asked, "What if we license their technology and integrate it into our product line?"

"What if cows had wings? We'd all have shit in our eyes," responded the general manager, immediately shutting down discussion. He didn't want to hear any new ideas or concepts that were other than what he wanted to hear. And he definitely didn't want any business assumptions challenged. This had a disastrous effect.

The research director had been hired by the previous GM. It was clear to him from this remark that he was not wanted. After a short stint in the job, the research director went to another company and then hired away the most talented members of the technology organization, much to the detriment of the business.

Use or allow *what if* challenges as a way of opening up the business to new ideas and showing that people are valued for creativity.

71. Eliciting Dissent

Agreement is not always necessary, nor is it desired in every business setting. People can and should disagree with decisions in a healthy business environment. Managers need to create an atmosphere where responsible dissent is encouraged. Many managers feel they do. However, the problem for a lot of managers is that they are completely unaware of the fact that they may be squelching dissent in spite of their efforts to encourage openness.

One of the nicest, smartest managers I have ever met directed a sales organization for a major industrial supplier of electronic components. His people all said very nice things about him, and by all accounts he was an ideal boss to work for. When he was transferred to run the marketing organization for the same company, he pulled out the old strategic plan, saw some clear holes in it, and developed a draft plan that he circulated for comment. His detailed plan was extremely complete and well thought out. He received no input.

He brought the plan to his VP, who sensed a problem when he asked about responses. He knew his troops well. "No comment" said "big trouble." He arranged for an out-of-the-office, all-day meeting for the marketing manager and his staff.

Yelling! Two straight hours of yelling at the marketing manager. It was unrelenting. They tore his plan apart, tore his management style apart—tore him a new orifice, as the saying goes. Why? What could he have possibly done to cause this? Nothing.

He had done nothing directly to his organization. He had also done so much that his staff felt there was nothing for them to do. And that's what caused him the problem. This was a guy who led from his head. He was smart, and he dominated his organization through the use of his intelligence. His thinking and his overwhelming command of information made it nearly impossible for anyone to have a thought of any value. The other problem with his style was that he was a "nice guy."

People didn't want to pick on him, so they just agreed with whatever he said and went on about their business. It never occurred to him that this attitude of his, his obsession for completeness and attention to detail, made dissent nearly impossible. Questions are the cure for this condition. When he received no response, he should have gone around and one by one asked people to share their thoughts. It may or may not have resulted in replies, but at least he would have been recognized as having had the courtesy of asking.

Use of *why* and *how* questions may elicit dissent:

Why is it important that we accept the decision?

Our time is short. How can we make sure that everyone who has an idea about this has voiced his or her opinion?

Why is no one willing to disagree?

Why do we not have at least three opinions for our course of action?

How could this be done differently?

How many different ways are there?

How can we do this for less?

Why do we do it this way all the time?

Why do we continue to rely on the same suppliers?

How is it that we continually find ourselves in second place?

How would you do this?

Why agree with the plan? If it were perfect, we would be the market leader.

I have uncertainties about the plan I drafted. How could we improve it? What changes do you think would help us?

These types of questions are also on the soft side. They don't ask "Who disagrees?" or "What's wrong here?" Employees are hesitant to respond with candor to these kinds of questions unless they happen to be in a very high-trust employment situation. Even then, the answers will be diplomatic because high-trust situations tend to breed collegiality and good nature.

Remember the chair-throwing engineers? Those guys would have no problem telling management what's wrong. Nor would they even be polite about it.

So, what happened after a whole two hours of yelling at the marketing manager? The guy just sat there and took it. He was shell-shocked. He had absolutely no idea that he was rolling over everyone with his logic, his command of market information, and his business acumen. Sometime during the second hour, his staff started to feel sorry for him. He was, after all, a nice guy just trying to improve the business. So they stopped.

His boss suggested that he take the elements of the plan, put them up on chart pads, and then ask everyone to pitch in and see what they could come up with as a team. The plan remained essentially the same but now included changes that were important to the staff.

Of course, there is also the opposite problem of intentionally quashing debate. I once worked for a manager who just didn't want to hear any dissent. He basically said this to his team by stopping all debate once he had made a decision on any particular topic. I stumbled upon this in my usual way—I disagreed with him, not realizing his management style.

I was newly assigned to his business at this particular moment, attending my first staff meeting when I blundered into a dispute. I had proposed a new marketing campaign. He didn't think it wise and prevented any discussion:

> **Manager:** I really don't like the idea of doing marketing to this market segment. It has been a market area where we have historically done poorly, and I believe it to be a waste of money.

But I blubbered on.

> **Me:** Well, I can understand why you might feel this way. However, we believe that we can significantly improve our market share in this segment by targeting the advertising directly to the potential customers rather than purchasing agents as we have done in the past.

A look of horror now appeared on everyone's faces. The "throat-slitting" motion was being directed at me by any number of staff members, as was the old "neck-tie hanging" sign. The boss then surprised everyone.

> **Manager:** You know, you may be right. What if we just scaled down the plan? Pick a few customer targets for a pilot study. Do you think that would work?

After we agreed to a small pilot budget of 10 percent of my original budget forecast, the boss turned around to his shocked staff.

> **Manager:** You know, you guys can disagree with me—just not all at once, and not all the time—but you had better do it with a good idea.

Although his last comment contained a little manipulative twist, it worked for a while. Everyone waited to see whether my plan would work, and it did. We increased sales into a market segment that the business had performed poorly in. When I asked for the original budget after the pilot worked, the boss said, "No." But the rest of the staff was now empowered to push back, from time to time. On the other hand, if my plan hadn't worked, it was very likely that I would have been transferred out of the division. Not only did he have a low threshold for dissent, he also did not tolerate failure.

If you do not seek dissent, it might seek you out anyway.

72. Are You Prepared for Any Answer? What About a Surprise?

CEO: Earl, can we count on you to reduce the cost of your operations by 20 percent?

Earl: If that's what you want, you'll have to do it without me.

CEO: Earl, did you just resign?

Earl: Yes.

The conversation is a short version of a longer story—a true one. No one was prepared for Earl to resign. He was the chief information officer of the corporation. I'm not certain whether Earl was prepared to resign before the discussion had taken place, but I am certain that the CEO was taken by complete surprise. Would you be prepared to deal with this kind of unexpected response?

Most of us are not. Here is a strategy, a plan for handling a surprise by employing a few questions to avoid making statements until you are certain about your response:

- **Reset the clock. Ask *what* questions.** Even if the answer was crystal clear, this strategy will reset the clock and allow you to absorb the answer a second time, when it's no longer a surprise.

  Q: What did you mean by that?

  Q: What did you say?

  Q: Will you repeat that?

  Had the CEO reset the clock in the preceding example, Earl might not have resigned. However, the CEO was caught so off guard that his next question was almost a reflex reaction. Earl may or may not have had a change of heart. But, backed into a corner by the follow-up resignation question, he might have felt he had no choice.

- **Engage in fact finding. Ask *how, what, when, where, who, how much* questions.** The key step when dealing with surprises is to look for facts that you need to respond appropriately. In the preceding example, no one knew what to do after Earl resigned.

  The CEO, an action-oriented person, went on by saying, "I'm sorry you feel that way, Earl, but you do understand the seriousness of our situation." The CEO went on the offensive. Many people use offense to defend against surprises. Resist this temptation, at least at this point, especially if you do not have all the facts of the situation in front of you. If you do have all the necessary facts, it's unlikely this is a true surprise.

Q: How did this happen?

Q: What was involved?

Q: Who else should be contacted?

Q: When did it happen?

Q: Where do we need to focus first?

Q: How much was ruined?

Q: How much damage was done?

Q: Was anyone hurt?

Also, as part of this line of questioning, avoid the *you* word: When did *you* decide to do this? How did *you* know it was ruined? Where were *you?* And so on. This might put the other person on the defensive. You need answers when surprises crop up, not confrontation.

- **Examine reasons. Ask *why* questions.** If you are dealing with a personal matter such as in the resignation just discussed, the line of questioning should probably continue in private. There are personal as well as professional reasons why a person might feel compelled to abruptly alter his or her career path. It may be necessary to insulate that person from view until he or she has a chance to explain.

  Our CEO in this example didn't do that; instead, he went on the offense. His reaction may have even hastened Earl's exit that same day. The company was left with organizational problems and lack-of-succession issues. Worse, other people started to follow Earls' lead. Look for reasons.

  Q: Why do you think you need to do this now?

  Q: Why was that path taken?

  Q: Why did this happen?

  Q: What were people saying before this occurred?

- **Draw conclusions for next steps, if any, and move on.** These few steps allow for a full discussion of the surprise, providing the manager with the ability to react quickly if necessary, or more thoughtfully if appropriate.

Could a surprise have been avoided if the CEO had asked Earl a different question? Perhaps. Managers are surprised all the time.

My first surprise in a supervisory capacity came to me in the form of a knife against my throat. It wasn't the first time a knife had been pulled on me while working at the steel mill, but I was very surprised at this move just the same.

Steel mills are tough places to work. The sign at our plant entrance announced the number of man-hours since the last on-the-job fatality. Every job in the mill was potentially personally harmful. Almost every kind of injury imaginable had occurred, from trains running over men working on the tracks to others being burned by molten steel.

Sporadic fighting also occurred, always away from the watchful eyes of supervision. There were shootings, knifings, and beatings with and without tools, plus the general pushing and shoving that might be expected in a schoolyard. The infirmary treated many men who "fell" while working.

I worked in the labor gang—that's what it was called in the mill. Few of the men had completed high school, and many of them had faced difficulties such as jail time. The men who worked in this gang were tough men for the toughest, riskiest jobs at the mill. I was dubbed the "college kid" after having worked there for a few summers between school years. I always let my beard grow to appear older and menacing.

The labor gang was responsible for the heavy lifting in the mill—we emptied boxcars using sledge hammers, cleaned out slag (waste material from the steel-making process), shoveled fly ash off girders to prevent buildings from collapsing, and cleaned out steel carriers with jackhammers. We did generally whatever it was that needed to be done.

On one particular evening as I was arriving for the overnight shift, I was met at the door to the locker room by the assistant superintendent (AS) who ran the mill at night.

AS: College kid?

Me: Yeah.

AS: You are the "pusher" on the overnight. Jimmy is out sick—might not be back for a while.

AS: Here's the list of jobs that need to get done—here's the list of your crew. Get the jobs done before 7.

That was it. This was the extent of my on-the-job training session. A "pusher" was generally a senior worker who worked directly under the foreman as a kind of supervisor whose job it was to push the men, push the work. Pushers were paid an hourly differential for assuming this responsibility.

It was my first night on. I had worked days and evening shifts for a while on the "cold" side of the mill—the place where the metal is rolled out, cut, polished up, and prepared for shipment to customers who turned it into refrigerator cabinets, washers, dryers, cars, and so on. I had been laid off from the mill three weeks earlier, and then, that day, I was called and asked whether I wanted to work overnight—the graveyard shift. So, I reported to the hot side. The money was better than any job I could have even dreamed of doing between semesters.

I called out the names of the crew and passed out the assignments—just like I had seen done on the two other shifts on the cold side. Only one problem: This was not the custom on the overnight shift.

Most of this crew worked two jobs. Jimmy, the current foreman, and all the foremen before him, had adopted a policy of allowing the crew to pick the jobs they wanted and sleep as long as they wanted. The only requirement was that the tasks had to be completed by morning.

I was the only person present in the locker room that night who did not know this. So, I passed out the jobs, turned around, and discovered a very impressive piece of highly polished chromium steel with an extremely sharp edge against my throat.

The mill was known as a place where people did get cut every once in a while. Industrial accidents happened all the time. This was the prelude to an accident.

Everyone else in the room appeared to be preoccupied with important tasks like tying their shoes. My mind immediately went into fact-finding mode.

Quite to my surprise, I heard myself say, "Is there a problem here that I am unaware of?"

"Yeah."

Great. It worked. Now all I had to do was guess the problem. "So, can you let me in on the problem?"

And so, the voice holding the knife explained how things worked on this shift. How they get their assignments done and then find places to sleep, holding just a few of the tasks undone until right before first shift so that the mill superintendent who arrived at about 6 a.m. every morning could see them working.

"Okay. So where do I sleep?"

Now they all started laughing. Evidently, the night foreman and pushers didn't get to sleep. Their job was to ride the trains around the yard with the night superintendent and prevent the mill super (the big boss), if he happened to show up, from discovering the sleeping crew. There was no way that the superintendent didn't know about the napping. It was a game.

So, just laughter, no cutting that night.

Would I have been cut if I didn't figure a way out? I have no idea. Like I said, the mill was a dangerous place. Although death was remote, I had seen a lot of fights and a lot of blood.

I trust that today's managers in most corporations are not faced with this kind of dilemma. However, there are parallels elsewhere to this kind of event. Employees may refuse to do the work assigned, or may call in sick on a critical workday, or may confront a manager with any number of creative surprises.

Surprises can come at you from any direction. Be prepared by using questions to help you manage an appropriate response.

73. The Use of Leading Questions

Leading questions are a specific type of trick question that suggest the answer, and they were discussed earlier as a type of question generally to be avoided. They do have a use, however, depending on what it is you are attempting to accomplish.

The manager in a previous example tried to convince the product manager that a new product was ready for market in spite of the fact that it wasn't. Normally, leading questions are not considered to be beneficial to a business discussion. They are often asked with the intent of trapping the respondent.

Q: So, tell me Mr. Jones, exactly when did you stop robbing banks?

Or they can be used to mislead a respondent to affect the outcome of market research.

Q: Would you choose the limited service of Bank A or the full service of Bank B, which has more locations, longer hours, and free checking?

Of course, I hope no one would ever ask such a skewed question in a legitimate market research survey, but there are many subtle forms of this kind of approach. The signal to the respondent is the use of qualifying words or expressions that appear in the question. Managers ought to avoid asking these kinds of questions in general discussions and hold them off until a time when their use is acceptable.

Leading questions may be used to move down a path to instruct or to develop a clear recognition of options to follow for the business.

For most business settings, a manager will want to form the question carefully; otherwise, respondents will feel manipulated. Before a leading question is asked, there might be a few preliminary questions containing plausible information. Or as in the case of teachers who ask leading questions all the time, the inquisitor is providing a lesson.

In the following example, the manager was not certain whether the appropriate customer support technical call center costs had been incorporated into the budget forecasts for the business. He also wanted to make certain that the information shared among geographically separated call centers was linked. Finally, it was a way of instructing.

A manager can ask leading questions provided that there is clarity in the objective being sought.

> **Q:** I take it that your budget request is complete?
> **A:** Yes.
> **Q:** Then, should we assume the 24x7 staffing coverage plan is ready to be deployed?

Leading questions are the least attractive kinds of questions to use in management settings, but they continue to be part of the repertoire of most managers. The suggestion to managers is to use them sparingly, and only after careful consideration of the objective of the question.

Too often, leading questions are used to trap a respondent in a particular answer.

74. Looking for Reasons

When you ask the *why* question, you are generally looking for something behind the words you see or hear. If you were a market researcher, vital information about a choice is communicated in the "Why did you choose the red dog?" question, for example. These questions are best employed when looking for reasons, rationale, or decision-making criteria:

What are your reasons for that?

I'm interested in your reasons for that. Can you explain it in more detail?

Why do you support that?

Why do you feel that way?

Why do you think this is the best course of action?

Why did you say that?

Why is this important?

Why don't our competitors do this?

Why do our competitors do this?

All *why* questions do not necessarily have to begin with *why*, nor do they all need to be in the form of an interrogative. The "I am interested in..." approach suggested above is an effective way of eliciting a response. It's still a good idea to add a brief question to go along with this comment, because the respondent may simply say, "Yes, I am interested in that, too," and you are left with the need to ask a question anyway.

These types of questions also encourage follow-up or probing lines of inquiry. Many managers accept whatever reasons are offered in the immediate response. This is an acceptable situation for short discussions, but if the reasoning behind a recommendation really needs to be understood, follow up until all reasons are exposed.

In addition, the use of the question *why* permits respondents to justify their actions or decisions. In many instances, managers may disagree with their colleagues, and this enables those with whom they disagree to explain themselves. If asked in a nonthreatening way, *why* is an open question that invites a responsible reply.

75. Are You Asking for an Opinion?

If your purpose is to solicit opinions, ask for them directly. If you are not the kind of manager who often asks for opinions, or you suspect people sandbag you, you might need a strategy to help you elicit them more effectively. This approach borrows from the way a lawyer might cross-examine an expert witness:

- **Try to establish some facts first.**

   Q: So, that 20 percent share of the market puts us in the lead?

   Q: The data is based on a test done in Arizona in winter. What about the data from Arizona in the summer?

- **Establish the reason you want the opinion.**

   Remember, your title is a license to ask questions, and it will still play a role in the kind of response given. Titles also place obligations on their holders to ask questions in a thoughtful way.

   Q: That's a remarkable achievement in so short a period. How long ago did you take over this product?

- **Ask for the opinion.**

   Q: This experience may be helpful. I would like your opinion. Do you think this same thing can be accomplished with the blue hair dye product?

A major clothing designer/producer sent her team of top managers to visit with one of their key suppliers. They were on a mission to find new materials to be used in a line of breakthrough athletic products. A host of new technologies were sorted through over a three-day time frame. Final selections were made after a series of additional meetings over the next few weeks, and a wrap-up session was held with the clothing company CEO.

After hearing the proposals, she turned to the key leaders of the team and asked for their opinion before making the final decision on new technology.

"I would like your opinion. Which of these do you think we should go with?"

The team all voiced their enthusiasm for a few of the options and then looked to her for the final answer.

"I don't like any of them." With that, she walked out of the room.

If you don't want opinions, don't ask for them. Otherwise, treat them with respect.

76. How Do You Evaluate New Ideas?

There are, of course, in the world a number of phenomena...that are just the result of general stupidity.

—Richard Feynman

Richard Feynman, one of the preeminent scientists of the twentieth century, recommended a formula for evaluating new ideas. Business, just like science or any other human endeavor, is a product of new ideas tried and tested over time. Some of these ideas are successful, and others are less so.

Some ideas are a reflection of a complete lack of good judgment. Here are some true-story examples:

- Taking your physically-out-of-shape deskbound business team out for a wilderness experience (Result: one death, one costly lawsuit)
- "Making it up in volume"—attempts to increase the sales volume of a product that cannot be produced for less than the price the customers are willing to pay (Result: eventual loss of the whole business)
- Protecting management pay and incentives while demanding cutbacks for employees (Result: resignations of management, death threats from employees, executives' cars ruined in parking lots)
- Acquisitions for synergy without making financial sense (Result: losses for shareholders and eventual divestiture)

Scientists tend to take a disciplined process approach to evaluating ideas. An analogue process would be helpful to the business community. When management relies only on whatever skills it happens to have, the results could be even worse than those noted above. Judgment is necessary but insufficient without a disciplined process for evaluating and reaching a conclusion. Subjecting ideas to a strategy to determine whether an idea is the best for the business given the circumstances of your business should improve ideas and their implementation.

Questioning New Ideas

- **Ask questions. What's the idea? (Explain it.)**
  Probe to show interest. What else do you need to understand?
  Avoid trick questions and derisive or sarcastic remarks.

- **What do you know?**
  Questions should be directed at finding what is true and certain.
  Ask questions from a number of different and relevant perspectives.

- **What are we uncertain about?**
  Q: What would it take for you to move from uncertain to certain?
  Q: What does your experience tell you?

159

Q: Have you seen this kind of thing before?

Q: What does the experience of the other members of your team tell you?

Q: How does the business handle uncertainty?

- **What is probable? (not what is possible)**

    Q: Is it reasonable?

    Q: Is it likely? How likely?

    Q: Is there any way to test the idea other than by fully implementing it?

A new idea is like a new product. It needs to be subjected to all the same kinds of scrutiny that you would exercise for a new car or new suit. You might not like it, but you really don't know much about it until you take it for a ride or try it on to see whether it fits.

77. Looking for Trouble?

Trouble occasionally bubbles to the surface in every organization, large or small. Asking questions always invites responses that may or may not have been intended. Management might go looking for trouble, or it may be revealed when unanticipated. In some situations, however, trouble of one kind or another might be just below the surface.

The point of the following inquiries described is that the questioner is intentionally looking for trouble. In these situations, how will you satisfy yourself that the problem solution process is effective for your needs?

- **Never accept a complaint as the problem. Ask more questions.**

    The higher in management you are, the more probing is necessary. Easy answers may reflect the problem you are supposed to see but not the root cause.

    Q: How does this show up in other businesses?

    Q: Why do you think this is the problem?

    Q: Can you stop complaining about the termites and explain to me why we have termites in a steel building with all plastic furniture?

- **Do not be satisfied until you see root causes (or probable causes).**

    Q: How do you define *trouble?*

    Q: What kind of "trouble" do we have?

- Use open questions until all facts, opinions, and options have been delivered.

  Q: Tell me more about....

  Q: That does appear to be the answer, but can you go back to the first time you noticed...?

  Q: Whom else should we speak with?

- Ask clarifying questions for a final focus.

  Q: So, that's how you discovered that our salespeople were selling the company automobiles? Tell me more.

  Q: Is there anything that has been left out?

  Q: Is there anything else you want to explain?

  Q: Whom else should we talk to?

- Ask concluding questions.

  Q: Are you satisfied?

  Q: Do you have any questions?

When you go looking for trouble, expect to find it. So, keep looking until you find all of it that you think there is—or all that you think you can stand.

78. Strategies for the Setting

Do you have a plan for specific types of questioning for certain settings, or do you just allow events to unfold?

The setting where a question is asked can be as formal as a stockholders meeting or as informal as sitting around a campfire. Depending on the setting you find yourself in, there may be different strategies to consider when asking questions.

Questions prepared in advance for staged settings—more formal settings such as meetings where expectations are clear about the kinds of things that are to be discussed—are very useful. For these types of situations, know your questions. It keeps the pace moving and will set a good example.

Staged or formal settings include the following:

- Debriefings
- Reviews

- Staff meetings
- Business meetings

Few managers I know prepare their questions for these settings. By training and experience, managers are well prepared to react by relying on whatever it was that propelled them into their positions. Preparation is left to the people who must present information. On one level, this works. However, this lack of preparation reinforces the role of management as reactive.

My contention, which is unscientifically based on a limited number of observations, is that mediocrity results from this process. I'm including myself in this criticism. Although I did prepare on occasion, I did so infrequently. But, I do know of one manager who prepared for all of her interactions.

She decided up front what issue was most important to concentrate on for a particular interaction and then charged ahead. This strategy worked as a hit-or-miss approach. When the information to be discussed in a meeting matched her preparation, the conversation was fluid and resulted in appropriate actions. However, as business conditions changed or information became available that did not match, the discussion that resulted from the interactions was like watching two ships pass in the night.

This often led to the inappropriate conclusion that she had personnel problems, because of the way in which some people answered questions. She was constantly moving people around to get a better fit—to get people who could understand what it was she wanted.

By the way, this didn't work. She eventually received the answers she looked for in every meeting, and her business unit never performed up to expectations. So, over preparing, thinking through every possible question, also involves thinking through the answers, which ultimately leads a manager astray. Remember, one of the purposes in asking a question is to find an answer—to find information that you don't know.

Most managers possess a basic skill set that allows them to consider what is before them and then manage this situation effectively. Some preparation would be helpful.

Some questions are appropriate for the setting, and some are not. The "Why can't we get/do/act..." and so on may help a manager in an informal setting broach a topic that she knows is a source of concern. If used properly, the question can be disarming and helpful to allow people to unload issues of concern.

Informal Settings

Type of Questions	Possible Approaches
Open	The "Tell me the story" approach.
Indirect	I wonder, *what* or *if* or *how.*
Rhetorical	Why us, why do we always get the tough jobs?
Hypothetical	*What if* (a good opener).
Negative	Why don't they get it? (a type of pressure relief)

Although the evidence suggests that there are no casual questions, there are casual settings. Managers would benefit much more by using a less-formal inquiry than by continuing the use of questions found more often in staff meetings. I know one corporate executive who is incapable of asking informal questions in any setting.

When he locks on to an issue, he pursues it without regard to where he is. It's difficult to enjoy a meal in his presence, let alone a cup of coffee. The problem is that he thinks he is having a casual conversation. If you have this problem, try to employ a different set of questions. They might provide the same information, or even more.

Be aware of the setting. Ask different kinds of questions in informal versus formal settings.

79. Are You Prepared for Answers?

You have heard a lot about questions. But what about answers? Responding to questions is a task all of us face at every level of organization, starting with ourselves, family, friends, businesses, boards—questions are everywhere.

When questioning improves, so do answers. It is the question that sets the standard for the reply. Remember the scribbled question in the corner of the magazine article? That was the level of reply being requested. Here are some simple rules to put into effect when answering good questions:

1. The answer should reflect the question (formal, informal, and so forth).
2. Speak (or write) clearly.

3. Project confidence.

4. If you don't know, say so—directly and clearly.

5. Avoid hyperbole and superlatives, such as *greatest* and *best,* and all-inclusive remarks, such as *always, everyone,* and *never.*

6. Be aware of your facial expressions and body language.

7. Think about the answer for at least a few milliseconds.

8. Know when to stop talking.

9. Avoid being too forceful or aggressive.

10. Avoid hostile questions. Don't take the bait.

11. Avoid being defensive, even if under personal attack.

12. Avoid personal attacks, slights, and belittling, demeaning, or personally harmful responses.

13. Stay away from the following types of answers:

    Misdirection (that is, answering another question rather than the one you were asked).

    Dead experts (quoting dead or unavailable experts unless you also quote available ones).

    Equivocating. Don't equivocate!

    Guessing. Avoid it. If you are guessing, say so.

    Incomplete answers.

    Do not violate the laws of gravity.

    Do not obfuscate. Leave that for diplomats and politicians.

    Do not throw red herrings, or any other colored herring. Do not throw any other foodstuff or any bull for that matter. There is plenty available everywhere else.

14. Learn to recognize when you have "made the sale." Stop talking!

The last recommendation works both ways. It's just as important to know when to stop answering a question as it is to know when to stop asking. So, if you have heard enough, you can stop the respondent.

Remember that the answer is a part of the question.

80. Are You Prepared for Nonanswers?

Nonanswers do occur every once in a while. The manager asks a question, and it is greeted by silence, or another form of complete disregard. The strategy is sometimes the old "buying time because you were not paying attention" type of response. It's a stall.

At a New York press conference early in my career, I was instructed to "stand in the back of the room in case I need to ask you a question" by our then executive vice president. The man was a brilliant scientist, a no-nonsense manager, and a steely-eyed leader.

My job was to provide information to the VP on my subject area of expertise should there be a question he was unable to answer. According to legend, in five years he had never called on anyone. He knew everything. Therefore, when he called on me, I had just enough time to cough out a powdered donut all over my dark blue pinstripe suit and ask rather boldly, "Could you repeat the question, please?"

There were about two hundred people in the room, and I wanted to make sure I was heard from all the way in the back. My philosophy was that if you are going to screw up, do it quickly, do it publicly, and take responsibility immediately. Then, go about your business, wiser for the experience.

My question was not really a question—it was a stall in the form of a question and quite legitimate under the right circumstances. It can buy you time to come up with a better answer than you might have given spontaneously. On other occasions, the question may have been garbled. In my case, I had no idea what the question was, let alone who even asked it. I was eating a powdered donut.

At the close of the conference, the VP approached the group of us gathered in the back of the room. Turning his back to me, he offered a helicopter ride back to the office to everyone he was now facing.

I was happy to ride the train home in silence. This incident did not weigh on my career, but it did weigh on my mind.

Your reaction to a nonanswer can say a lot about your management style and can be very instructive to the nonrespondent if handled appropriately. (I have remained attentive in every meeting, conference, and discussion I have participated in since that time.)

81. Have You Asked About the Fatal Flaw?

Consider this scene. A senior management review of a very successful business unit has just been conducted. Everything is working well. Revenues are rising, profits are up, and costs are down. The relatively new CEO of the company is asking his final questions of the chief technology officer (CTO).

CEO: Is there a fatal flaw in this business, in our technology, or in our product line that worries you?

CTO: There is no fatal flaw.

CEO: Are you certain?

CTO: Yes.

CEO: Why are you certain? Is there nothing that keeps you up at night? No problem you worry about? Is there one thing that if it were to happen would destroy the business, or is there one thing that must happen in order for your business to be the success you have forecast it to be?

CTO: Nothing I can think of. I sleep well.

One month after this business review, a meeting was held between the chief technology officer of this company and the chief technology officer of the single largest customer of the business. This was not really unusual because both companies conducted a fair amount of business together in many markets where they were both buyer and seller of raw materials and finished product.

Other CTO: We have just learned that one of the materials we both use to produce our product line might be a potentially serious environmental concern. Although we have no evidence of a direct health threat or proof of environmental damage, yet, so serious do we view this problem that we have decided to stop production immediately and go out of the business of selling products that contain this material.

Company CTO: Are you serious?

Other CTO: Yes. We just thought it fair to inform you since we purchase large quantities from you. We also supplement our process by producing small quantities, too. We were unable to find a substitute material anywhere in the world and felt certain that you must also be aware of this problem. We're hoping that you have a technical solution to the problem. If not, we will sell out the inventory and stop production.[31]

The flaw that no one saw suddenly became fatal. It was not many things. It was one thing. It was one, unanticipated, unintended, unaccounted-for detail that struck down the business. CEOs do not normally go around looking for fatal flaws in successfully operating businesses because these enterprises generally don't have them. But, new CEOs do (or should). It keeps them up at night until through questioning and experience they learn all the flaws of their business and, if one is found to be fatal, they take the steps necessary to correct it if possible. In the preceding case, the problem could not be corrected.

The fatal flaw question is different from any other. It is designed to expose a hidden problem such as a faulty assumption or inappropriate conclusion. It has a number of side benefits, too: It requires deep thought about the answer, it challenges people, and it requires commitment.

A fatal flaw is one thing—not many problems, it is one. It is the one thing that, if it happens or does not happen, could cause the demise of the product, the business, or plan all by itself. Conversely, it could also be the one thing necessary for the business to succeed. In either case, it is the proverbial "all your eggs in one basket" dilemma.

Here are some real examples of not having asked this question.

Disposable Diapers

A picture appeared in the *New York Times* some years ago showing an artist's conception of a mountain of toxic diaper waste wrapped in indestructible little packages stacked on a Staten Island landfill. It dwarfed the tallest buildings in the distance. This picture represented the growing environmental concern over the use of nondegradable materials by the disposable diaper industry. In this case, the "joke" was that nondegradable materials were being used to package the most degradable substance personally produced by man.

A concern over the environmental hazards of nondegradable waste was growing, and diapers, adult incontinence pads, and other personal-care products were portrayed as the potential villains.

The environmental movement had heightened everyone's awareness of the issues. Legislatures started to hold hearings, and entrepreneurs envisioned major business opportunities based on degradable materials.

A race started among a dozen leading companies to see who could come up with degradable materials for diapers. The motivating factor was that the entire industry would be forced to adopt them by regulatory or legislative

bodies. High-technology materials companies, egged on by the diaper manufacturers with potential promises of purchases in the billions, collectively
poured hundreds of millions of dollars into the opportunity. These new materials, however, were costly to produce.

"What if the cost of the diaper would have to increase by 30 percent? Who
could argue with the law? The consumers will have to adjust to the higher
prices." That was the unspoken, but very real rationale of many of the companies. Companies were betting millions of dollars on legislatures composed
of...elected officials.

Now what politician worth his or her election return is going to legislate
higher costs for young families—voters? What politician is going to legislate
higher costs for the elderly who use incontinence pads and for all other consumers of bed pads and tampons—more votes? Not one. Yet, this was the
only way that the products from most of these competitive companies would
be financially attractive.

After spending literally hundreds of millions of dollars in R&D, most of the
materials technology companies abandoned their efforts or converted them
into other, smaller opportunities.

The diaper companies responded to the needs of the environment as well as
their consumers without being required by law. They and their current suppliers did understand the environmental concerns and the competitive pressures from degradable materials. Why were the material manufacturers
unable to figure it out?

They neglected to ask the fatal flaw question. Here is the discussion that
never took place:

Q: What one thing is required to happen in order for this diaper material
business to be a success or could ruin this business if it does not
happen?

A: We need an act of Congress to force all manufacturers to adopt
biodegradable materials.

Q: The chance of Congress passing legislation forcing companies to
raise prices on diapers?

A: Not very high.

Q: Why are we going to spend millions of dollars developing this
technology?[32]

The one thing that needed to happen for this to be a financially successful business was that the material had to be cost-effective. However, the cost of the material, even at the best manufacturing company at that time, was higher than the price of the material that diaper manufacturers were currently purchasing. Therefore, the only thing that could make the new materials financially attractive was for biodegradable materials to be required by law.[33]

Build It—And No One Will Come

A major plastic producer developed a new material. It had a number of unique attributes that made it attractive. They had made plenty of material in the laboratory to test the potential uses of the material, and all tests showed that it could be manufactured at a lower cost than many products that it would compete with. The company decided to build a new plant to manufacture the plastic.

They had plenty of experience building large-scale manufacturing facilities and so decided to skip the scale-up pilot plant step. Manufacturers often build smaller versions of large manufacturing plants to make sure the process works in large-scale operations and to learn about any unanticipated problems. This step helps identify problems and may be done a year or two ahead of building a large facility.

Years of experience had taught this company how to get things done right, the first time. However, they were also pressured by cost problems and time constraints. In addition, market pressure was building on their old product lines, and profitability appeared unsustainable. The faster this new manufacturing facility could be built, the better off the business would be.

In this case, all the right questions were asked. A lot of time was spent on the "What could go wrong?" question. Contingency plans were made for each of the possible problems, and the facility was built.

What was the fatal flaw question (that was not asked)?

Q: What one thing could ruin this business?

A: If the technology is not scalable.

That's exactly what happened. The plant that was built was dead on arrival—a total loss for the business. What worked on a small scale in the laboratory, and then in a very small-scale facility, had not worked in a large-scale operation. It never operated, and the entire mess (plant, equipment, and technology) was later sold off. The business failed.

Hair Today—Gone Tomorrow

A popular hair stylist built a small successful business in the Midwest. He leveraged his appeal by opening three additional salons that caught the attention of investors, one of whom was a customer. A plan was developed, and money was poured in to expand the chain into dozens of salons in the region and then take it national. It failed, and much money was lost. Here is the conversation that didn't happen:

Investors: When did you have your last physical exam?

Señor Maurice: I've never had one. It's against my constitution.

You can figure out the rest of the story. He was in poor health, although he looked great. He died shortly after a lot of money was spent on locations, staffing, training, and the development of marketing campaigns of which he was the star. Even entrepreneurs can get caught without thinking through the obvious questions that should be considered. In this case, Maurice was relatively young and healthy and had an infectious zest for life. Most investors believe they have enough savvy to know what to ask and when to ask it. And, most of the time, they are correct. But, they do make mistakes, like most of the rest of us.

The Doctor's Plan Was DOA

While we are on the subject of investors, a final fatal flaw story comes from the experience of a sophisticated East Coast venture capital firm. Once again, no names.

The firm was approached by a physician, Dr. Z, who had a business idea and new technology in the form of pending patents. His concept was well thought out, and the physician himself had made significant personal investments in pursuing patents and furthering the technology to the best of his ability. He had "skin in the game," as many investors like to say. His plan to build a business based on his technology made a great deal of sense.

A decision was made to invest money in developing the business. A research facility had been set up to move the technology forward, people were hired, and a business team was put in place. The plan was dead on arrival—at a very unusual location.

The managing partner of the venture firm happened to be traveling to a meeting on the West Coast, where he ran into an old friend of his sitting at an airport coffee shop with a third party—another physician in the same field as Dr. Z. So, the partner causally asked him whether he knew Dr. Z. He was not prepared for what he heard.

"Yes, I do!" Then, Dr. Y started to discuss Dr. Z's technology in great detail as Dr. Z had described it to him a year or so earlier at one of those small, invitation-only scientific meetings (held before a patent application had been submitted). Because the meeting was a small gathering and proceedings were not published, Dr. Z must not have considered this a public disclosure and forgot to mention it. Not only had Dr. Z discussed the technology, he had also handed out a small write-up with some data to explain it to his colleagues. Dr. Y himself had started to develop additional ideas for which he too was considering for patents.[34] End of story.

The fatal flaw question should be asked of every plan, every new technology, every new service, every product, and every opportunity. This is especially necessary for businesses that invest a large amount of capital in facilities, equipment, and human resources.

Listening

I know that you believe you understand what you think I said, but I'm not sure you realize that what you heard is not what I meant.
—Richard M. Nixon[35]

82. Listening: The "Hearing Phenomenon"

The other half of asking a question is how it is received and perceived, and whether it has had the intended impact. Answers alone are not a full indication that you have communicated effectively. There are two additional factors in this equation: how the question was heard, and then what you do with the answer. We discuss how you heard the answer in a moment. What was heard is the key factor in this section, and managers are sometimes completely unaware of how a question is heard.

They tend to think that just because a question was answered, it was heard correctly—that the respondent listened. But here is a very real discussion I had with an associate (A) in a large accounting firm.

A: My boss grills me every morning when she gets in.

Me: What do you mean she "grills" you?

A: I get in early every day, and when she arrives, she always walks by my desk and asks questions.

Me: Like what?

A: Like, "You are going to be fired today!"

Me: That's not a question.

A: Yeah, but that's what I hear every time she asks me one.

He is a young person who has been with his firm for a couple of years and gets good reviews. Yet he hears this every time his boss questions him, whether intended by the boss or not. I have discovered that this "hearing phenomenon" is not uncommon, particularly in service businesses. Checking into a major name-brand hotel at four in the morning, the desk manager told me she hears "you're second-rate" every time her boss questions her.

In reality, these people are not facing imminent firing; they are listening through an amazing array of filters. People hear all kinds of things from their managers, particularly when asked questions. They hear things such as "You're doing a good job," You're doing a bad job," or "Your shoes are too pointy." So, listening to the question followed up by listening to the answer is always needed.

By the way, our observations on questions/answers, and particularly listening, extend to e-mail communications—these should be included in your thinking, too. More and more manager–employee interaction is taking place electronically rather than in person. As managers become remote from the place where the work of employees is actually conducted, understanding is becoming even more important to the process of asking a question.

Before going further, let's get some definitions out of the way. There are two basic concepts just described: hearing and listening. In spite of the preceding example of hearing something the manager did not explicitly state (we have no idea whether the accounting manager was harboring the thought of actually firing the employee, nor do we know whether his paranoia was justified), this activity still comes under the category of "listening." It may be thought of as creative listening, but it is listening just the same.

- *Hearing* is the physical response of the body (your ears) to sound. Your ear hears the sounds whether you listen to them or not. Hearing may be completely active (as when one is attending a concert) or passive (where the person may or may not focus on content) or, in many other cases, a combination of both. It is also possible to hear without listening; there are a number of excellent counseling books on this subject.

- *Listening* is a conscious recognition of the sounds you hear. It is active, and it is a skill. The intent of the listener is to understand what is being said, what was meant by it, and then to determine the appropriate response, if any.

The distinction for managers is that hearing (Dad, did you hear anything I said?) requires you to receive the spoken word while listening (Yes, I heard you say you were just about to practice your violin in spite of the beautiful rainbow you were describing that appears so close that you could reach out and touch it) and applying your skill as a manager to the content of what is being spoken.

We hear and listen quite naturally. You don't need this book to instruct you on how to do either of these things. What we can do here is point out some of the pitfalls to avoid and opportunities that a manager can gain by employing a few listening strategies. Some managers, even those who are very good at asking questions, may be notoriously poor examples as listeners.

You can find many good books on listening offered by psychologists, clerics, teachers, psychologists, and musicians. I'm not going to repeat any of their fine advice here. They have three basic messages in common: Listening improves communications; listening improves your enjoyment of many things, such as music; and listening is a way to learn, thus making you better at your chosen profession. A number of interesting scholarly papers have also appeared, with listening being studied as a skill that can be taught[36] as well as a physiologic phenomena that can be examined medically.[37] Studies will continue to improve our understanding of listening as a skill, but what do we do in the meantime?

You should consider four questions with respect to listening as the other half of the questions:

- Are you heard?
- Do people listen?
- Do you listen to yourself?
- How do you know you are understood?

What people hear is not always what you intended to communicate. The only way to be certain of their understanding is to listen to their responses to your questions.

83. What Are You Listening For?

A business manager I knew communicated almost exclusively by voicemail at times when he knew most of his employees were unavailable or were not answering their phones.

Me: Cal, why do you always communicate by voicemail?

Cal: I'm not interested in a quick answer—just the best answer.

Me: Why not just send a message?

Cal: Voices are better for me. I can tell a lot about the person as well as his or her answer. Are they feeling well; do I hear stress in their voice or maybe disgust? I'm responsible for people. Without them, we have no business.

Cal was a professional listener in every sense of the word. He was the best listening manager I have ever come across. His questions would occasionally miss the point, but he was clearly someone who was much more interested in the full substance of the response, not just in the content. He liked to "hear" the response as well as listen to it. That was his philosophy.

The downside of relying too much on hearing and not enough on listening to the content is that we all tend to listen for what we expect to hear.

Cal occasionally missed things because he tended to hear what he was selectively listening for. So, when he heard trouble in the voice of his distribution manager, he generally responded to the symptom as opposed to digging in to find the source of the problem. Perhaps he did not want to know—I never asked him that—or perhaps he did know and was in denial. Whatever the case, without using some questioning tools to augment great listening skills, it's possible to miss important issues.

This is how one of his trusted managers ended up "redirecting product" to generate a huge sum of money to cover the financial shortcomings in his salary. Some probing questions with follow-up strategies might have made the guy squirm enough to come forward with his misdeeds before the audit uncovered the wrongdoing.

One way to think about avoiding the problem of becoming a habit listener is to consider this headline posted as a CNN report in 2001:

Listening for Secret Nukes, Hearing Giant Meteors[38]

Listening stations have been operating for decades all over the world. The listeners are predisposed to hear certain items of interest. They use this approach as a filter for weeding out information of interest from communications that are of no consequence to them. In this case, their attentiveness to listening for the secret explosions of nuclear devices enabled them to hear meteors crashing into the atmosphere long before any reports were made by "more public" agencies.

The sounds of the answer you get to any question you ask may be signals for you that contain information of importance that is not in the content. Once again, in the preceding case, Cal missed the meteor—missing product—because he did not match the content (come on up and spend the weekend on my yacht) with the stress in the manager's voice when they discussed sales shortcomings.

If you are going to listen, you should be hearing everything.

84. Avoiding Listening Errors

As managers with ever-growing responsibilities, busy schedules, and varied commitments, we are all subject to a number of listening errors. Most of these are easily corrected provided that you are aware of them. This is a short list of the most serious kinds of listening shortcomings:

- **Interrupting.** Interjecting before the respondent has had a chance to fully answer the question.
- **Ignoring the answer.** Behaving as if you asked a question just to hear yourself talk as opposed to listening to the answer.
- **Acting distracted.** Packing for a trip while conducting an interview.
- **Walking away.** Hard to believe, but I have seen managers ask questions and literally wander away during the answer.
- **Repeating the question.** Losing your train of thought so completely during the answer that you need to ask the same question over again.
- **Misinterpreting the response.** Thinking that the respondent agrees with what you have been saying just because he or she answered your question.

These six errors can be divided between *common sense* (ignoring, repeating, and misinterpreting) and *common courtesy* (interrupting, walking away, and acting distracted). It is common sense to simply pay attention to the answer to a question you have asked. Doing so might require concentration for someone who is easily distracted by the swarm of bees that may be swirling about his or her head, but for most non-bee-threatened managers, all that must be done is to focus.

One of the main reasons I have found managers not paying attention to the answer is that they are considering what to ask or say next. It doesn't matter. If you don't pay attention to what respondents say and how they say it, they will start to ignore you, too.

In an office of a U.S. company with business in China, a manager had placed signs around the office with transliteration of Chinese words so that the staff might be alert to them if used in conversation with people who were still attempting to master English. Her boss arrived at the office one day and asked what she was doing to improve communications between her staff and their Chinese counterparts. She explained the signs to her senior manager. Moments later, during a tour of the office, the senior officer asked her, "What the heck are these signs for?"

This boss might have been having a bad day, her ulcer may have been acting up, or perhaps she was just getting over a nasty head cold—whatever the reason, she did not listen to the answer to her own question. She had merely asked it for effect and then proceeded on with her agenda because she probably had no expectation that the manager was addressing the problem in a creative way. This is bad enough, and likely to result in jokes about the boss, but errors of courtesy are much worse than this shortcoming in my opinion.

A manager who interrupts his or her employees during their response to questions is showing an enormous amount of disrespect to them, and to himself or herself as well. Not only does this devalue the person speaking, it also diminishes the manager in the eyes of all present. Many successful people have developed this habit. I have seen managers as high as CEOs interrupt people in the middle of answering questions during discussions with them. They would likely not tolerate the behavior in return.

Walking away needs little explanation. I have witnessed this only once and understood it to be symptomatic of a brilliant person who had an attention-deficit disorder. Still, it was terribly troubling and inexcusable.

A final bit of insensitive managerial behavior was exhibited by an HR manager interviewing a prospective employee while packing a briefcase. It's no wonder why the employee did not accept the position offered.

> You might not remember what it was you asked, but people will remember that you listened.
>
>

Conclusions

85. Is Socrates to Blame?

Use of the Socratic method[39] improves critical thinking. Socrates, who lived from 469 to 399 B.C.E., employed many different types of questions as a method of developing within respondents conclusions they would otherwise not have drawn. This "method" has been a staple of the teaching profession ever since.

Raise your hand if you know who Anytus was. If you are also familiar with Meletus and Lycon, you get extra credit. These are the three guys who brought legal proceedings against Socrates. They were responsible for the death of a 71-year-old man because they viewed him as a threat. Why? Because he asked questions! These three guys had all the answers. The last thing they wanted was questions.

Socrates was forced to drink a poison because, among other crimes, he was found guilty of corrupting the youth of Athens. He generally did this by showing them the folly of their elders! His questions and his method of asking (and his manners) were often unsociable—he relished the opportunity to make an ass of people, particularly of people he did not like or whom he felt were inferior (which meant everybody). His questions were often embarrassing and rude, even for the standards of ancient Greece, where even the gods could behave in an undignified and unsociable manner. However, he was, nonetheless, a threat because he asked questions.

Socrates challenged people constantly by peppering them with questions of logic, of critical thought. He probed, asking follow-up after follow-up question until his targets would surrender, not because he forced them to accept his position but because he was able to show them how inane their beliefs were.

These three guys—Anytus, Meletus, and Lycon—thought they had all the answers (and Socrates, all the questions). They thought that life would improve in Athens without all his questioning. Just how successful were these guys? Well, look around. Does Zeus still rule the world? Can you still settle a debt by sacrificing a chicken or a gerbil?

Socrates' questions were intended as tests that no one could really pass. His desire was to demonstrate to those in positions of knowledge, power, or authority that no matter how much they thought they knew, they were ignorant and that it was much better for everyone involved in their enterprise to admit their ignorance right up front.

"Ignorance for failing the tests," as Socrates might say.

The Socratic Manager

The Socratic manager asks questions proceeding from what is known to what is not; it's a forward-moving query. To paraphrase his method: Seek always to establish where you are now in knowledge—your understanding of markets, programs, marketing, sales, operations, and your business in general.

Socratic management is based on what is known as Socratic irony. It starts with a profession of ignorance. Socrates, for instance, might have managers attend a meeting and probe for where a lack of knowledge exists. Then, instead of the usual moment of revealed wisdom that strikes managers in retreat types of meetings, there might be a moment of revealed ignorance. From this profession of ignorance stems his method of questioning.

The mission of Socrates was to rid people of the illusion of knowledge—there was always something that they did not know, even about an area in which they were considered expert.

Many Managers Have Been Given Horses to Manage

A horse is a liability to a person who tries to manage it without having enough knowledge.

—Socrates

The knowledge gained by managing in one situation in one particular business may not be transportable to the next one, particularly if the manager proceeds from the premise that useful knowledge was gained in the last assignment.

Perhaps it would be much better to begin from the premise that you have a horse and that you don't know much about managing horses. Therefore, it would be useful to start by questioning rather than any other course of action. Got a new horse? Start asking questions all over again. Every horse is different (different personality, different breed, and so on).

Socrates always requested frankness whenever he was asking questions. "Do be frank about answering whatever I may ask of you." Although this is often expected in business, it's not always the case.

The Socratic manager might, for instance, view presentations as inquiries rather than as discussions. Too often, I have watched a senior manager, a leader in the business, page through a copy of a presentation inquiring not one bit into any of the cogent details of the arguments included, but merely looking at the results—the financial forecasts—and asking questions about these outcomes as if they were real.

Management Wisdom

Every man is wise only in respect of that which he knows.

—Socrates

Although Socrates worked to find a basis for ignorance, Taylor[40] worked from a basis of knowledge. In either case, it is necessary to proceed forward by asking questions. To dispel the notion that knowledge makes a person wise, the following conversation was recorded by Plato, between Socrates and Euthydemus.

S: Tell me, do you think a wise man is wise in relation to what he knows, or are some people wise in relation to what they do not know?

E: Obviously, they are wise in relation to what they know. How could anyone be wise in relation to what he does not know?

S: Then, are they wise because of their knowledge?

E: What else could make them wise?

S: And do you think that wisdom is anything other than what makes people wise?

E: No, I do not.

S: So, wisdom is knowledge.

E: So it seems to me.

S: Do you think it is possible for a human being to know everything there is?

E: No, indeed.

S: So, it isn't possible for a human being to be wise in respect of everything?

E: No, certainly not.

S: Then, every wise man is wise only in respect of that which he knows.

E: So it (now) seems to me.

And so it was that Socrates proved to Euthydemus that we are all unwise, all ignorant in that which we do not know in spite of the fact that there are many people who are considered wise. Socrates would have us believe that these people are wise because they ask questions rather than have answers, because one cannot have all the answers.

The same things [knowledge] are assets if one knows how to make use of them, and they are not assets if one doesn't.
—Socrates[41]

Make better use of your assets. Ask questions.

86. Conclusions and Final Recommendations

My premise is that asking questions is a discipline that managers need to learn. Some people are naturals, but even they need to think about improving or adding to the techniques that come easily to them. The rest of us must work at it.

I once attended a corporate meeting of business leaders from a large multinational major company. It was convened to address a serious problem in the company. Everyone was expressing concern about the lack of information. As the meeting wore on, it became clear that a consensus was building around the need to hire a firm to "go find answers for us."

Bullshit!

What?

Bullshit! We have all the answers we need right in front of us.

This time it was not me irritating everyone in the room, but one of the older directors—a guy "on his way out," so to speak. He was bullet-proof. He could say and do almost anything short of the unethical, immoral, or illegal, and he was still going to retire soon with a full pension.

He went on to explain:

> We have all the answers. This company employs tens of thousands of people all over the world. There isn't anything about the business that we do not understand. How can there be? Just get out there and ask for it! You will get all the answers you want—and quick, too.

Just ask. That is the simple conclusion of this book. *Just ask!*

Of course, you need to know what to ask, how to ask it, of whom and under the right circumstances, and so on. "Just asking" will get "just answers." What you want are the answers you need to improve the business, solve the problem, or develop a new idea.

To put this simple singular conclusion into effect, I suggest these simple guidelines in summary:

1. **Enter all situations thinking about what you don't know.**
2. **Consider all others with whom you speak as equals.** They might not have your title or responsibility, but they know what you don't, and they are probably expert at it. You need them.
3. **Be yourself.** Don't fake a question or adopt a style because of this text or any other. Be yourself. If a style or type of question doesn't fit you, choose one that does.
4. **Always thank people for their answers.** Do so even if the questioning was contentious or particularly difficult. This will go further in establishing your legitimacy as a manager—as a leader—than your skill as an interrogator.

No footnotes appear on the bottom line of an annual report explaining to the reader that good questions were asked that revealed great answers and improved business performance. People might not even remember who asked the right questions. Over time, however, the business will perform better, with better questions from all levels of management.

All businesses are operated by people. You can interrogate a search engine all day long, but the machine is still not as supple as the human mind. All the answers you need to improve your business are there. Just ask.

Epilogue

Are You Still Here?

So, did I survive my first year of punishment in the corporate world? Yes.

After nearly a year riding airplanes every day, I returned to the home office to demand a better job. My personal record of plane rides was 22 in an 8-day period (lots of short hops and before the days of tight airport security). When I returned home from this exhausting week, I sat sideways on a chair, leaning into the corner of the kitchen...only to wake up there 9 hours later. It was time for a change.

I walked into my manager's office prepared to issue an ultimatum: Another assignment or I was history. My travel time had allowed me to check out market opportunities, so I realized that I did have value elsewhere if the company was no longer interested in me. I didn't even have to raise the issue.

He greeted me warmly, as if I were a returning war hero and offered me a new assignment along with a nice promotion. This made me feel very confident in myself. Oh, was I still very naive.

I assumed that my retention and promotion were due to the great job I was doing—that my management finally realized the attributes I brought to the job. Perhaps I was also forgiven for whatever transgressions I may have been guilty of when I started (transgressions such as telling the truth). Was I ever wrong? You bet.

Unbeknown to me, an important customer, our largest customer in California, had written the CEO extolling my virtues and mentioning that she would not buy any product unless it had my personal approval. When I visited with her to examine a product-quality problem, I had told her to toss out all of her inventory because much of it appeared to have been damaged in some way. Then, I yelled at people in the plant to have new product shipped out immediately.

The reason for my promotion was explained to me on a trip to a business meeting a few months later. It was while traveling with a senior vice president that he explained my real employment situation. The customer letter was the reason for my success. Period. In a way, it was because of my hard work, but hard work that was valued by a customer. My management was another matter. He told me that I had really pissed off my local management and that I was never likely to have their support again.

He could see that I felt a little deflated. However, what I had done, he went on to explain, was to draw a line in the sand, and on one side I put the best interests of the company. I then refused to cross it even at the risk of my employment. This had impressed senior management, and I was placed on a list of people who were considered to be "corporate promotable"—people who senior managers mentored as potential candidates for future leadership positions. This was not a guarantee of future success, but it gave me a positive perspective of the company.

I might have lost the confidence of immediate management on the way out the door, but I walked back in with the confidence of our customers, the most valuable commodity in any business.

My new assignment was to work on a new product. The difference this time was that I was assigned at the beginning of the development cycle and not at the end. This would give me a chance to work through the entire product development process. I was very excited.

The product manager called a meeting for the next morning. When I walked into the room, I was greeted by a familiar voice, one I had not heard in almost a year.

"Are you still here? I thought I told you to leave," snorted the manufacturing superintendent, without even looking up from his coffee.

Without saying a word, I turned and left the room. But I did not leave the company...just the assignment.

Definitions

Definitions are offered to clarify the skill of asking questions. The objective is to remove doubt about the intent of any of these words found in the text. The words used in this short dictionary and in the text constitute a type of jargon around the skill of asking questions, using common words.

abuse of managerial power   Use of a managerial or supervisory position to ask a question that would be generally unacceptable among equals. (Can you lie to me about the forecast?)

avoidance   The practice of ignoring questions in spite of the fact that they might be required by the situation. (Sales are down, but volume is up.)

casual question   A question asked among family, friends, peers, or acquaintances. There is no such thing as a casual business question asked by a manager.

clarifying question   A closed question aimed at defining specific information.

closed question   An interrogative usually of the *what, who, where*, and *when* variety that requires a specific answer. It is the type of question most often used in examinations (cross-examinations and probes).

compound or nested question   One question containing two or more points that need to be addressed in the answer.

constrictive question   A question that cuts off discussion or implies that the conversation should be terminated.

context   The environment that provides a questioner and the respondent with a general understanding of the use of the question and the answer.

convergent question   A closed question designed to focus on reducing the scope of a discussion. Convergent questions can be filtering, clarifying, or either/or types of questions, to name a few.

187

defensive qualifier   A phrase defending the need for the question, or excusing the questioner for asking it, immediately before asking.

direct question   A clear, unambiguous, understandable interrogative focused on a specific respondent and generally of the *who, what, when, where, how,* and *how much* variety.

divergent question   Usually asked as an open question, or even a hypothetical question, it is an interrogative asked for the purpose of expanding a discussion beyond the current boundaries.

double-direct question   A compound question containing two items that can be responded to even though the interrogative is about one of them. It is usually asked as a leading question.

dual-answer question   A closed question where a yes or no answer might mean the same thing.

fatal flaw   The one thing that if it were to happen or if it were not to happen is either required for success of the business or would spell doom. A business usually has only one fatal flaw, if it has any.

filtering question   An interrogative asked to specifically exclude information from the answer.

habit question   A favorite question or the question (or questions) asked by a manager so often that most people are aware of what will be asked before hearing the question.

hypothetical question   Asked often in the form of *what if* or *suppose,* it is generally an open question used to diverge in a discussion.

indirect question   The *who* of this interrogative is unclear, as is sometimes the subject being spoken to. These are usually in the form of *why, I wonder,* and *tell me.*

jargon   Language that requires translation to be understood by others. It is usually equivocating in some way.

leading question   An interrogative designed to offer a conclusion or answer in the question for a respondent to follow.

loaded question   An interrogative that includes a presumption or offers an assumption that may have negative implications for the respondent.

modern management   A style of management that was founded by Frederick Winslow Taylor. Managers, according to Taylor's definition, are "nonproducers."

negative question   An interrogative asked with negative wording such as *why can't we*, or a question that has a negative pretense.

negative nested questions   Negative and positive questions asked together in such a way as to render the answer meaningless in advance.

neglected question   An unasked question, also sometimes known as a *pregnant question.*

nested question   See *compound question.*

normalization of a defect   A problem that appears so often that it is considered normal.

objective   A goal that is measurable and can be reached in a finite period of time.

open question   An interrogative that does not limit the response to a single type of answer. It permits the greatest latitude in an answer.

OTE   Overtaken by events. Can also be referred to as "OBE"—overcome by events.

positioning   That part of a question, usually the preamble, which defines the inquisitor.

posturing   That part of a question, usually the preamble, in which the inquisitor tries to appear superior to the respondent.

prejudicial question   A question that contains a biased opinion or remark.

provocative question   A challenge issued to the respondent(s).

question   A word, comment, phrase, facial expression, or physical gesture intended to elicit an expected response. It is an interrogative interaction.

questioners   The individual asking the question, the interrogator. There are many different types of questioners. Here are some of the primary types of questioners: examiner (and cross-examiner), explorer, analyzer, inquisitor, inspector, scrutinizer, checker, interrogator, pollster, researcher, hunter, interpellator, interviewer, tester, indagator, groper, ransacker, grabbler, and questioner.

redirecting question   An interrogative that returns a respondent to a previous question, statement, or issue.

respondents   Respondents may be considered to be examinee, replier, answerer, or relater.

rhetorical question Although asked in the form of a question, a rhetorical question is intended to be a statement, asked in an interrogative format for effect.

right question A question that is appropriate for the circumstance and that produces the desired results.

spinning A form of intellectual flatulence whereby a question is answered in a way that alters reality to suggest that the respondent's characterization should be accepted when, in fact, it resembles a view of the world more closely held by dung beetles in a cow pasture—meaning that the answer that is spun is pure crap.

Socratic management A business-operating style in which managers work from the premise that they are ignorant.

Socratic method Asking questions as the way in which a person (a teacher, for example) enables another to draw his or her own conclusions, or to learn what it is the teacher (Socrates) is attempting to communicate. It is a "do not tell, ask" approach.

strategy A plan for reaching a specific objective.

stupid question There is no such thing as a stupid question.

References

American Management Association. "Strategy: How can I become a more critical thinker and increase my productivity?" www.amanet.org/askama/strategy1.htm.

Baldoni, John. "Are You Asking the Right Questions? To get the answers you want and push your agenda forward, you need to know what the right questions are and when to ask them." Harvard Management Communication Letter, Harvard Business School Publishing, 2003.

Barg, Gary. *The Only Stupid Question Is the Unasked One.* Caregiver.com, 2002. www.caregiver.com/editorial/stupid_question.htm.

Beckwith, Harry. *Selling the Invisible.* Business Plus, 1997.

Bergson, Lisa. "The One Question I Didn't Ask." *Business Week Online,* May 19, 2003. www.businessweek.com/smallb12/content/may2003/sb20030519_7090_sb002.htm.

Bertlein, Barbara. "Asking Questions." *Business Journal of Milwaukee.* May 5, 2003. www.//milwaukee.bizjournals.commilwaukee/stories/2003/05/05/smallb5.html.

Bossidy, Larry and Ram Charan. *Execution, the Discipline of Getting Things Done.* Crown Business, 2002.

Browne, M. Neil and S. M. Keeley. *Asking the Right Questions. A Guide to Critical Thinking* (8th ed.). Prentice Hall, 2006.

Burton, Gideon O. *Silva Rhetoricae* (The Forest of Rhetoric). 1996–2003. http://rhetoric.byu.edu.

Courtney, Hugh. *2/20 Foresight.* Harvard Business School Press, 2001.

Day, George S. and Paul J. Shoemaker. *Wharton on Managing Emerging Technologies.* New York: John Wiley & Sons, 2000.

Dontonio, Marylou and Paul C. Bessenherz. *Learning to Question, Questioning to Learn.* Needham Heights, MA: Allyn & Bacon, 2001.

Doyle, Sir A. C. *The Complete Sherlock Holmes.* Golden City, NJ: Doubleday & Co., 1927.

Einstein, Albert. *Relativity: The Special and General Theory* (Masterpiece Science ed.), translated by Robert W. Lawson. New York: Bonanza Books, 2005.

Farrall, Stephen, et al. *Open and Closed Questions.* University of Surrey, 1997. www.soc.surrey.ac.uk/sru/sru17/html.

Feynman, Richard P. *The Meaning of It All.* Reading, MA: Helix Books, 1998.

Finlayson Andrew. *Questions That Work,* Amacon, NY: American Management Association, 2001.

Gilroy, John. *Basic Neurology* (3rd ed.). New York: McGraw Hill, 2000.

Hamel, G. and C. K. Prahalad. "Strategy as Stretch and Leverage." *Harvard Business Review,* March–April 1993.

Haydock, R. and J. Sonsteng. *Advocacy, Examining Witnesses: Direct, Cross, and Expert Examination.* New York: West Publishing Co.,

Katzenback, J. R. and D. K. Smith. "The Discipline of Teams." *Harvard Business Review,* March–April, 1993.

Leeds, Dorothy. *Smart Questions.* Berkley Books, 1987.

Levitt, Theodore. *Thinking About Management.* New York: Free Press, 1991.

Lindstrom, Martin. "The Art of Asking the Right Questions." January 21, 2003. www.clickz.com/brand/brand_mkt/article.php/1571531.

MacFarland, Jennifer. *Leadership and Learning: The Art of Asking Questions.* Harvard Update. Harvard Business School Publishing, 2001.

Martel, Myles. *Fire Away!* New York: McGraw-Hill/Irwin, 1994.

Martin, Jim. *Interview with Chris Clark-Epstein, author, 78 Important Questions Every Leader Should Ask and Answer.* www.unisys.com/exec-mag/strategy/internbal/leadership/2002_12dialog.htm.

Mauet, Thomas A. *Trial Techniques.* Aspen Law & Business, 2002.

Mcteer, Robert D. "The Dismal Science? Hardly!" *The Wall Street Journal,* A16, Wednesday, June 4, 2003.

Meyer, Christopher and Stan Davis. *It's Alive.* New York: Crown Business, 2002.

Moore, David P. *The Little Black Book of Psychiatry.* Malden, MA: Blackwell Series Ltd., 2000.

Nunberg, Geoffrey. "Initiating Mission-Critical Jargon Reduction, Ideas and Trends, Talking the Talk." *New York Times,* WK 5, August 3, 2003.

Pappas, Marjore L. and Ann E. Tepe. *Pathways to Knowledge: An Inquiry into Learning.* Libraries Unlimited, Teacher Idea Press, 2002.

Payne, Stanley L. *The Art of Asking Questions.* Princeton, NJ: Princeton University Press, 1955.

Saunders, Trevor (Ed.). *Plato: Early Socratic Dialogues.* Penguin Classics, 1987.

Sobczak, Art. *Using Trial Lawyer Techniques in Sales.* The Sales Crusader, 2001. http://sales.crusader.hypermart.net/a_question.htm.

Stone, I. F. *The Trial of Socrates.* New York: Doubleday, 1989.

Taylor, William. "Message and Muscle: An Interview with Swatch Titan Nicolas Hayek." *Harvard Business Review,* March–April, 1993.

Washington, Tom. *Interview Power.* Mt. Vernon Press, 2000.

Watkins, Jane Magruder and Richard J. Mohr. *Appreciative Inquiry: Change at the Speed of Imagination.* John Wiley & Sons, 2001.

Wellman, Francis L. *The Art of Cross Examination* (4th ed.). Barnes and Noble Books, 1992.

Whiteley, Alma. *The PATOP Model for Developing Managers' Critical Thinking/Questioning Skills.* Faculty of Education Language and Community Services, 2001. www.//ultibase.rmt.edu.au/Articles/June97/whitel.htm.

William, Kenneth B. et al. "The Art of Asking: Teaching Through Questioning." *Academic Radiology* 9, 2002, 1419–1422.

www.lib.msu.edu/digital/vincent/findaids/Watergate.html.

www.watergate.info/chronology/1973.shtml.

Questioning as a Spectator Sport: Where to Go to Watch and Learn the Game

The popularity of news interview shows and, of course, the ever-present game shows where contestants/celebrities must answer questions to win prizes (usually money) continue unabated on television. We participate by listening passively to all of these formats—interviews, news shows, television court, sports stories, and myriad other conversational offerings—so many that it is difficult to keep up with them all.

Television offers us an opportunity to learn from professional questioners, inquisitors, and interrogators. They can be viewed for their use of strategies, how their questions are composed, and the approaches they take in employing strategies. Because it is TV, you can observe their body language and facial expressions, and match them with the tone of the person asking.

Some of the people who conduct interviews or pose questions on inquiry-based news and investigational story shows can teach us about questions by example. Many are very good—they have to be just to capture enough viewers to entice sponsors to fund their broadcasts. A small number stand out as excellent examples that managers can learn from.

Four criteria are suggested for you to use in the evaluation of professional questioners:

1. The use of rules for disciplined approaches to questioning.
2. They employ multiple types of questions in each interaction.
3. The use of strategies as opposed to a script for asking questions. (They might, in fact, be working from a script, but their questioning appears natural and to be following a clear path.)
4. They appear always on the lookout for the "fatal flaw" in the discussion.

My personal list of people whom I believe are the best I have recently seen in this field is noted in the following table. I constructed a short list based on these criteria. That doesn't mean that others are not equally good or even better interrogators than those on my list. Also, note that their presence on the list does not mean that I agree or disagree with their style, intent, or the views they may imply or express. It simply means that, in my opinion, they have mastered the art and the science of questioning.

Every one of the people on this list has a different style, but they all have one skill in common: They are expert in using questions as tools. They are not enamored of one particular kind of question or with one specific approach. Their job is to get answers, expose a story, or to learn something of interest from another person or situation. This is not unlike the job that managers face.

Best Questioners	Key Attributes
Barbara Walters	Direct, open, and closed question strategies
Ted Koppel	Direct, follow-up, use of multiple strategies
Greta Van Susteren	Redirects and probes
David Letterman	Question combinations, expressions, body language
Bryant Gumbel	Intensely purposeful, focused, no escape questioning
Geraldo Rivera	Double-direct, great strategies for closed questions
Bob Costas	Open questions that are as good as closed, focuses on story
Jerry Seinfeld	Questions that expose how we think critically
Jon Stewart	Socratic

You will notice the presence of a comedian on this list. Many comedians use questions as the setup for their humor—for their punch lines. Why? Because the question gets the audience to participate by thinking of the answer, allowing the comedian, of course, to deliver an answer that is usually unexpected. In other words, you must answer; otherwise, there is no humor. You might think comedians are just up in front of audiences entertaining with jokes, but many of them are performing questions. In my opinion, Jerry Seinfeld does this better than anyone in the business. But that's my opinion.

Many other professional questioners are equally good for managers to learn from, but they employ a limited number of techniques—some of them only a single approach. They are known by this approach, and it can be instructive for managers to consider when and where their techniques may work. I am not advocating that their styles be adopted but that you look past the form and into the substance of their skills.

Questioner	Questions and Strategies
Al Franken	Questions as commentaries, open but closed
Bill O'Reilley	Leading questions, challenges
Sean Hannity	Direct negative questions, and nested negatives
Larry King	Indirect, casual approach, in a staged setting
Larry Kane	A Philadelphia news fixture, totally answer focused

There are many others for whom questioning is an art as well as a science. Many of the local media outlets around the country also have good interviewers who can serve as instructors for anyone wishing to examine their approaches.

No answers, just the questions please,
Then all will be revealed.
You must behave like Socrates,
Just to earn a meal.

Endnotes

[1] What if cows had wings? The *what if* question is the only one on the list that allows a manager to step out of his or her formal role and engage people in a unique way because it suspends the normal rules of managing.

[2] Fred Taylor studied shovels, and he studied the men who wielded shovels. He concluded that large men can use larger shovels and smaller men need smaller shovels. That is not management. That is observing and explaining the obvious. The management part of his story came next. He concluded that the smaller man needed a shovel small enough to enable him to shovel fast enough to shovel the same amount of material as a large man with a large shovel. So, the manager's job was to match the man with the right shovel. That is the foundation of modern management. Fred Taylor was hailed by Congress as a genius.

[3] The company in this story produced medical products. It has been bought and sold since the time of this story.

[4] The product manager was eventually promoted, perhaps more slowly than she would have been if all had gone well. Years later, she eventually left the company to pursue other interests.

[5] Here are two websites that you can visit for more information and to listen to Sam Ervin and other Watergate figures: www.satergate.infor/chronolgy/1973.shtml and www.lib.msu.edu/digital/vincent/findaids/Watergate.html.

[6] Fred (not his real name) did so well running this customer service operation for a technical service company that he took a job leading customer service for a consumer products company. He later became an executive with a brand name retail chain and still sees "daylight" all 24 hours of the day in his operations. By the way, he did not consider his original assignment as manager of the night and weekend crew as a particularly good

career move for him. As a matter of fact, although he was well liked by management, they thought he had "limited potential." (Ha! Ha!) I happened to run into him at an airport one day and asked him about his success. He mentioned that his worst assignment turned out to be the best one of his career. He also attributed his success to a skill he picked up working in customer service: good listening. You cannot be successful in that line of work without it. He said he just started "listening to the young kids," and if it sounded like a sound business idea, he followed their lead.

7 The actual study by Deloitte Consulting, by Brian Fugere, found that "The three year growth rate of straight talking companies was better than those of companies that obscured their communications with baffling verbal fog." The quote is from an article in the *Telegraph* on June 24, 2003, http://www.telegraph.co.uk/core/Content/displayPrintable.jhtml.

8 The pass could be intercepted, the pass could be incomplete (thus wasting a down), and the pass could be completed. To Woody Hayes, a completed pass was almost as good as a gain on the ground, but not quite.

9 The decision could turn out to be wrong. The decision could have no effect and, therefore, look wrong because the manager wasted time on it. And finally, the decision could actually be the right one, but someone else will take credit for it—they always do.

10 One other management lesson can be taken from this example: A business cannot have two number one priority projects at the same time. Management must decide and then must staff the one considered most important with their best people. Sharing personnel across projects confuses the teams by creating personal challenges, as cited in this example. The rationale used by the business was that the second "top priority" project was to enter the market ahead of the one that was identified as most important to the future of the business. Everyone was aware of this. Neither product was particularly successful.

11 This term was noted by the *The Wall Street Journal* in a front-page article describing the shuttle investigation at NASA. (May 22, 2003)

12 Security and Exchange Commission—the mission posted on their website: "The mission of the U.S. Securities and Exchange Commission is to protect investors, maintain fair, orderly, and efficient markets, and facilitate capital formation." You can understand how damaging this kind of managerial behavior can be to a company as a prelude to misleading people in an irresponsible and potentially damaging way.

¹³ Pick any one. Rehab is all about dealing with the reality that a person has been denying by use of a bad habit.

¹⁴Vulcans, although incapable of lying, are indeed capable of deceptive behavior, and Androids can be programmed to lie. So, the only place you should do business with them is a at a Star Trek convention.

¹⁵"Are You Asking the Right Questions?" John Baldoni, Harvard Management Communication Letter, March 2003, Article reprint No. C0303C.

¹⁶The expression *high-impact words* is from Haydock and Sonsteng, in their book guiding lawyers in developing cross-examination skills, *Examining Witnesses: Direct, cross, and expert examination (Advocacy)* (NY: West Publishing Co., 1994).

¹⁷Quoted from *Leadership,* by Rudolph W. Giuliani.

¹⁸One excellent book on critical thinking is *Asking the Right Questions: A guide to critical thinking,* by M. Neil Browne and Stuart M. Keeley (Saddle River, NJ: Pearson, 2004). It describes a method for improving thought processes.

¹⁹ *Ibid.*

²⁰ *Profiting from Uncertainty: Strategies for succeeding no matter what the future brings,* by Paul J. H. Schoemaker (NY: Free Press, 2002).

²¹ Francis Wellman in the *Art of Cross-Examination* (Touchstone, 1997).

²² *Hannity and Colmes* was a "liberal versus conservative issue format" show in a kind of yin and yang tradition on Fox News. It may or may not be on the air when you read this. Just think of any television news interviewer raising a finger to constantly paint a point in the air and you have the picture.

²³Richard Feynmann: "I wonder why I wonder why I wonder why I wonder."

²⁴Sherlock Holmes to Dr. Watson.

²⁵I am using *grandstanding* to mean two things; first, the deliberate capture of management attention by one or more persons who desire to be noticed; and second is the filibuster by one or two persons who want to exert some control over the discussion to serve a different agenda.

²⁶Double-direct questions are drawn from the legal textbook *Examining witnesses: Direct, cross, and expert examination (Advocacy),* (NY: West Publishing Co., 2004), by Haydock and Sonsteng.

[27] This question is often cited as a leading question, and many leading questions are trick questions. They are designed to yield information that the person might otherwise not provide.

[28] For more information on follow-up questions from the perspective of an interviewer, refer to Payne, who discusses some of the questions noted here (as well as others).

[29] Sherlock Holmes to Dr. Watson.

[30] Avoid taking the bait by following up on this misdirected answer. This is also a signal to probe—see this same situation under the section "Follow-Ups and Probes."

[31] Do you ever wonder why some good products just disappear from the store shelves, never again to be found? This is one case that resulted in such an outcome. The finished product made by the company posed no known threat, but the production caused one of the key ingredients to "escape" into the environment. It was not known to be toxic, but the long-term effects of exposure to the material were unknown. Because production facilities were in the United States, Europe, and China, the potential was great enough to cause one of the two companies to take the unprecedented step of going out of the business of producing that particular product. It was a serious blow to both companies.

[32] During the frenzy of degradable development activity, I had the opportunity to meet with a VP of a major European diaper and incontinence product supplier. I asked him how much of a market did he envision once all technical difficulties with degradable materials had been worked out. None. No market opportunity would emerge. So, when I challenged him as to why his company continued to encourage the makers of degradable materials to provide him with sample products (thus implying there was an opportunity), he retorted that there was one need his company wanted to meet: supplying adult incontinence products to government-operated elder-care facilities in the one country in the European Union that required degradable materials. That was the only opportunity in the world where there was both a requirement that degradable materials be sought and that the buyer be willing to absorb some cost increase.

[33] The story of how one company turned this fatal business flaw into an opportunity is told in the book *Radical Innovations: How mature companies can outsmart upstarts*, by Richard Leifer et al. (Harvard Business School Press, 2000). A team of faculty from Rensselaer worked with a number of companies to develop a perspective on how large corporations

were handling innovative opportunities. Check out the story of Biomax (hydrobiodegradeable polyester).

[34] Although most businesspeople understand this concept, it is not general knowledge that all new inventions need to be formally protected before they are discussed in public with anyone. We can debate whether the invitation-only scientific meeting constituted a public disclosure; Dr. Z obviously did not think so. When an invention is exposed publicly by any means, the inventor has one year to file a patent application in the United States. However, the opportunity to gain patents outside the United States is generally lost. So, the objective in every case is to guard inventions as closely as possible until the appropriate paperwork has been filed. In this example, the physician really believed that the private meeting was, in fact, confidential because all attendees were asked, informally, to keep them as such. Dr. Y did not seem to feel the need to avoid discussing the technology, as scientists focus on advancing knowledge, and this is best accomplished through papers and discussion. It would be up to the lawyers to figure out the appropriate legal standing of the technology—the business decision needed no such effort. All investment in the idea ceased following the coffee conversation. In this case, the fatal-flaw question was asked and answered before the investment. But given the naiveté of the inventor, more probing was required to enable him to remember.

[35] Richard Milhous Nixon was the thirty-seventh president of the United States. His quotes are preserved for history on a number of different websites. All presidents end up with a series of less-than-flattering quotes because their every public utterance is captured for posterity. Here is a sampling of some of his other remarks: "If you think the United States has stood still, who built the largest shopping center in the world?" and "Solutions are not the answers." For amusing quotations from notable personalities, go to www.brainyquote.com. It posts quotations from Democrats as well as Republicans and a variety of other nonpolitical personalities.

[36] Kubota Shinya, Norio Mishna, and Shoji Nagata, "A study of the effects of active listening on listening attitudes of middle managers," *Journal of Occupational Health*: 46, 2004, 60–67.

[37] Stephen M. Wilson et al., Listening to speech activates motor areas involved in speech production, *Nature Neuroscience* 7:7, July 2004.

[38] "Listening for Secret Nukes, Hearing Giant Meteors," by Richard Stenger, 5/23/01, www.archives.cnn.com/2001/tech/space/05/23/secrete.meteors/index.html. The human ear cannot detect the low-

frequency infrasonic waves emitted by meteors hitting our atmosphere or by nuclear weapons exploded in secret places. Intelligence agencies hear a lot of these kinds of sounds and keep it to themselves unless the "noise" becomes public. I wonder what else they hear?

[39] This is a method of inquiry in which questions are used as the primary tool for investigating an idea, advancing a thought, or winning an argument. Plato positions Socrates as the "curious questioner" in his stories about him. Wikipedia is the primary source of this information. There is also a very good book, *Trial of Socrates*, by I. F. Stone that examines how grating Socrates' method of asking question after question was on his fellow citizens, particularly because he was challenging their morals. And, of course, Plato wrote a number of Dialogues recording the discourses of his mentor.

[40] You still remember Frederick Taylor—the guy who built the modern science of management based on shoveling.

[41] "...unless he sells them." Critobulus in *The Estate Manager*.

FT Press

FINANCIAL TIMES

In an increasingly competitive world, it is quality
of thinking that gives an edge—an idea that opens new
doors, a technique that solves a problem, or an insight
that simply helps make sense of it all.

We work with leading authors in the various arenas
of business and finance to bring cutting-edge thinking
and best-learning practices to a global market.

It is our goal to create world-class print publications
and electronic products that give readers
knowledge and understanding that can then be
applied, whether studying or at work.

To find out more about our business
products, you can visit us at www.ftpress.com.